I0702161

The Series,,, 'TheresMyMind'

A visual storyteller using funny photography is the motivation behind these books. . Come enjoy the hilarious stories and offbeat photos of a storytelling photographer mixed with many true and factual details. The goal is to capture your attention and transport you to another world.

The Website: https://theresmymind.com/

Come explore the chaotic beauty of life as my quirky brain sees it. I never know where life will take me next, and since my brain is firmly attached inside my head, I never know where my mind will be either. But, I haven't lost it yet, hence the name "there's my mind", a whimsical space dedicated to visual storytelling and hilariously inspired <u>funny photography</u>.

The Blog: https://theresmymind.com/travel-photography-cool-literature/

Raising Awareness about world travelers, unique photography, great literature.

PUBLISHED BOOKS:

- **Where is My Mind**

- **What is Electricity?** *and where does it go When it leaves the toaster?*

- **Art Imitating Life,** *and Other Observations*

- **Doors & More...** *The Magic of San Miguel de Allende, Mexico*

- **Magical San Miguel de Allende, Mexico,,,** *Its Doors & More*

 **Illustrator:**
Noel RedBuffalo
BRADENTON FL.
NoelRedBuffalo@redbuffalotattoos.com

Editor: Margaret Henkel

Photographer: Ed Kramer

Ed Kramer

ed@ejkramer.com

First edition April 2024

ISBN: 979-8-9857234-8-9

Dedicated To:

Margaret Henkel, my creative graphic designer and life partner who has given me the motivation to complete this story,,,

LET THE *SPARKS* FLY!

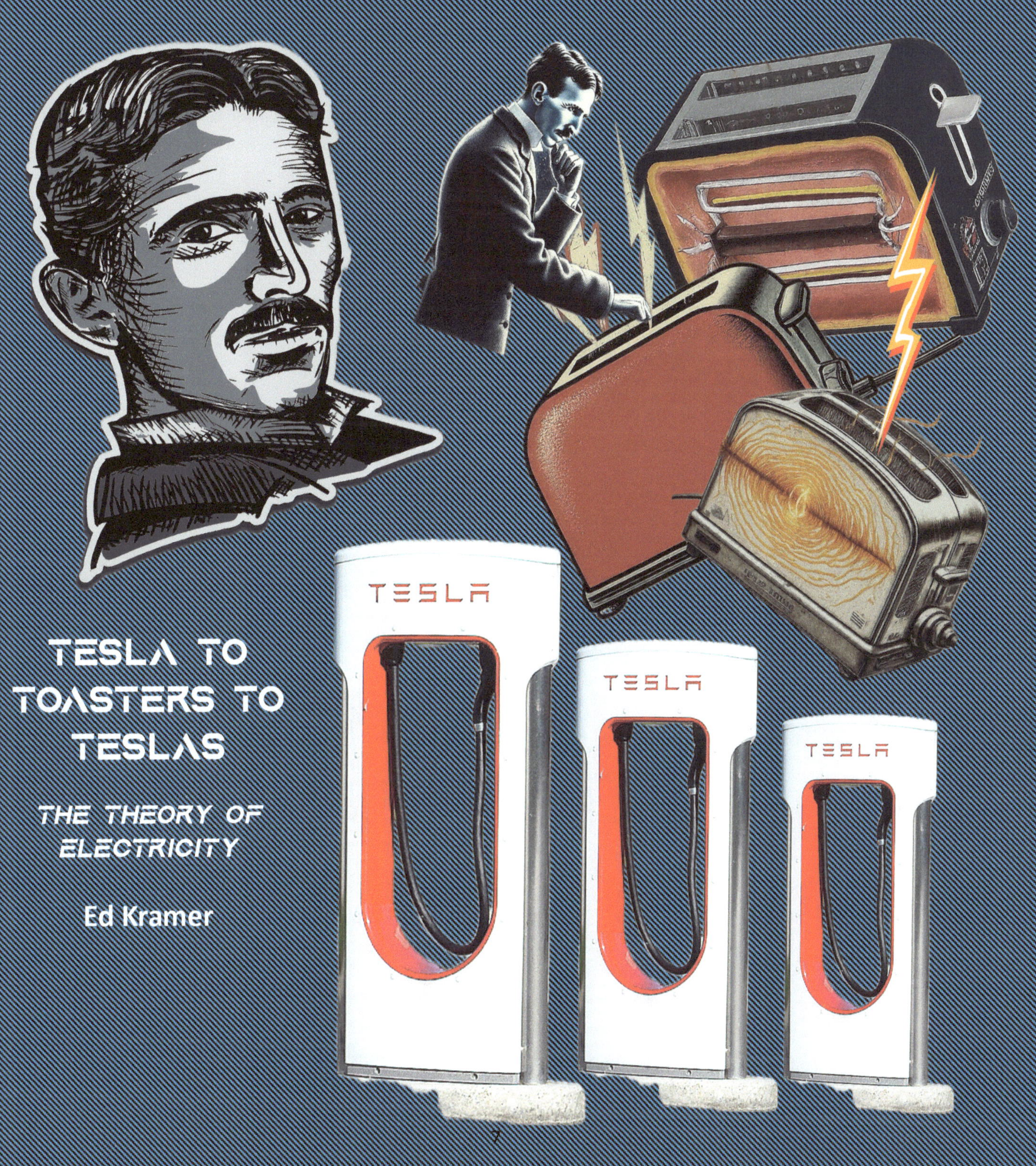
TESLA
TESLA
TESLA
TESLA TO
TOASTERS TO
TESLAS
THE THEORY OF
ELECTRICITY
Ed Kramer

THEORY OF ELECTRICITY

T O E = *Th*eory *Of* *E*lectricity or *Th*eory *Of* *E*verything

This book has a great amount authenticated detail,
awesome wit and factual references,
a lot of photos, and with portions bordering
on a deep dive into disinformation.

As a matter of national security — I have to share this total story!

Eddie Current

Stop that immediately

shouted Kramer, Senior as he raced across the room toward me, sparks shot out of the receptacle and the lights went out in the entire house. You guessed it, my interest and infatuation with electricity, began at a very early age. This was partially encouraged by my father who was an electrical engineer and was working for a municipal electric power company.

As I grew up, birthday and Christmas gifts at an early age involved electrical products, but focused on those that had educational value. Many of the gifts came from a company called HealthKit. At the time, they made the world's finest kits for radios, amplifiers, and shortwave, where you can build professional quality electronics equipment yourself, even if you have little or no experience, but as my father raced across the room, the lights went out, he stumbled on a chair and swore to teach me a lesson I would never forget.

My knowledge and interest in electricity took off like a rocket headed toward Mars.

In the first chapter, you will be introduced to Eddie Current, who will be your guide thorough this complex maze. starting with the Theory Of Electricity, and then on to the challenges of electricity and Climate Change.

The purpose of this book is to have some fun, laugh at some of the absurdities of the climatologists and our governmental bodies, and to gain a bit of knowledge about our powerful energy sources and uses.

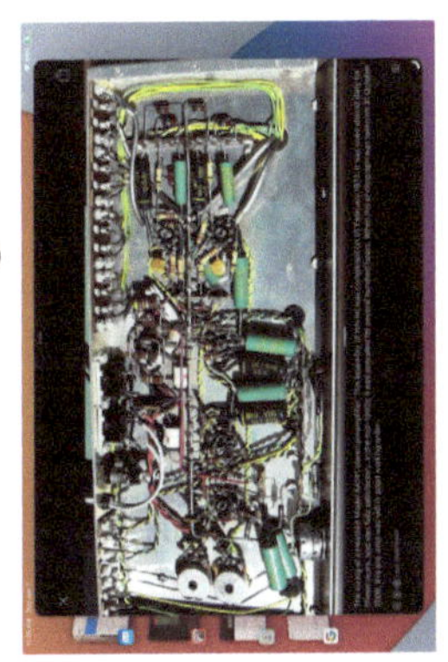

Lets start by introducing Eddie Current.

Growing up is a tough situation for electrons.

Eddie Current at an early age of about 8 told his dad, Mr. Kilowatt, that he wanted to be a pirate when he grew up.

Dad was quite surprised 'cause most young electrons at that age wanted to be cowboys or firemen.

Thank goodness dad did not take him seriously and set a schedule for peg leg surgery and an eye removal!

As he grew up, he found enjoyment in comic books, mostly those involving Super Heroes.

Early childhood toys,,,

Eddie Current's favorite flavor of ice cream is shock-a-lot

But becoming a **Super Hero** was not his calling. If I could become one, I would have to have the power of empathy. Empathy allows us to understand and feel the experiences of others, which empowers us to take action and make a positive difference in someone's life. He imagined that if he all had the power of empathy, he would be able to truly listen to and understand without judgment or bias. He could use his power to break down barriers and create connections between people of different backgrounds, cultures, and beliefs. We could use it to promote kindness, understanding, and compassion throughout the world.

This seemed to be too limiting to him, and his skills and ambitions were maturing.

On a trip to Las Vegas he experienced the Sphere if Influence,,, and was inspired. This book is Eddie Current's Sphere of Influence. Eddie Current found his calling and was appointed by the government as
Ambassador of Power.

So let's tap into our own superpowers and embrace empathy. Let's use our unique abilities to make a difference in the lives of those around us and work towards creating a more empathetic and connected world. Together, we can make a powerful impact on the world and create a brighter future for all.

If you wanted to become a motivational speaker or writer, I believe that each and every one of us has the potential to make a positive impact on the world, no matter how big or small. And while superpowers may seem like something out of comic books, they actually serve as a great metaphor for the power we all possess within ourselves.

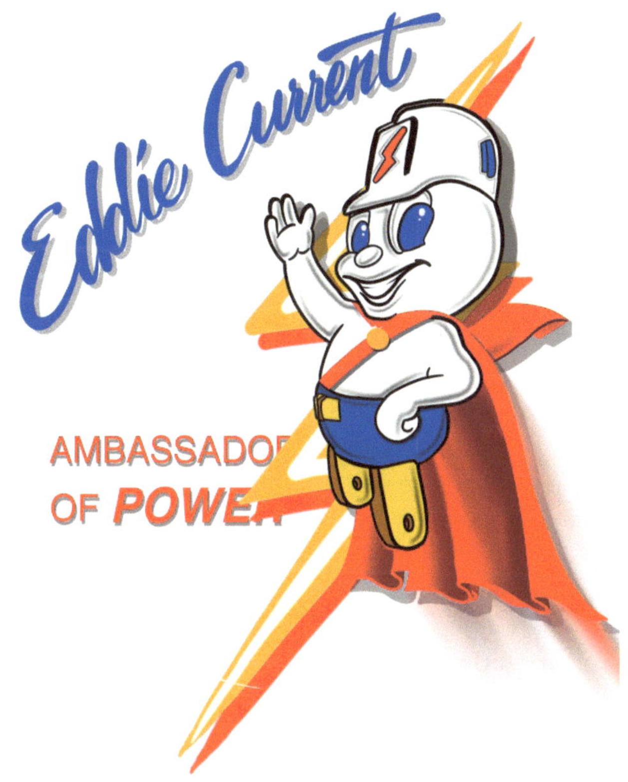

Understanding Electricity is complex and most readers find it hard to follow all the details. But the underlying ideas can be clearer than the details. Information is an important tool, but with a bit of humor interjected, the reading goes much faster. So, my goal to give you an explanation of some of the tasks that are critically dependent on electricity as well as an explanation of many of our every day encounters with this mystery of this magic.

Where to start? Some folks suspect that energy like money flows in and out of your life, and others suspect that it only flows out. Today, energy and its generation are major conversation points in every discussion when the topic gets to climate control. We often hear the terms green energy, climate change, humans are the main cause of climate change and the creation of green house gasses.

Hopefully, we will demonstrate the inter-relationship of these commonly used terms. Understanding energy, in this book, is focused exclusively on electricity. The first book describing electricity is titled *What Is Electricity,,, And Where Does It Go When It Leaves The Toaster*. You met **Eddie Current**, *Ambassador of Power*. Eddie has awaken, and like with Rip van Winkle, who wakes 30 years later, sees the whole scene of electrical power and its distribution has changed. These chapters help bring us up to speed starting with Tesla, to Toasters, and back to Teslas.

Are you ready for a thorough explanation of the legends and the mis-information that has inundated our lives and our dependency on this miracle.

We are going to squeeze as much as possible out of Eddie Current, so fasten your seat-belt and start flipping the pages.

Lets clear the air. If you are a legalist, then you might think that this is an eddy current.

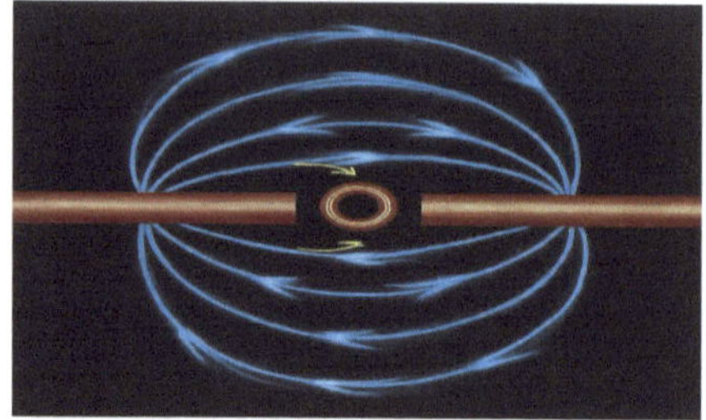

But it is not! He is actually a multi-skilled fella doing the many tasks we require of him. Sucking up the darkness in our homes, powering our TVs and music systems and laptops, charging our smart phones, cooking our breakfasts lunches and dinners, powering our EV automobiles, and we are just getting started.

On the following pages, you will discover several of the Eddie Current specialists. Maybe they are like doctors. Some are internists, others might specialize in specific fields, like neurosurgery, pulmonary, orthopedics, pediatrics or gastroenterology. Do you think it is the same electrons that turn the wheels of your electric vehicle also power your smart phone? Do you think that they all come from the same source? Are the electrons that come from coal and oil power plants the same as the ones that come from the wind. What about the Eddie Currents that come from solar panels? What about the Eddie Currents that come with the ever-ready batteries you buy in the store, or the lithium batteries you have in your smart phone?

Maybe electrons look like a glass of orange juice, or a bottle of Mountain Dew. I am confident we all have referred to electron flow as turning on the juice!

What is an Eddie Current? He's not a person. He's not an Animal. He doesn't have a gender or a nationally. No one knows what he looks like. What is a kilowatt, or a kilowatt-hour? Can you get a bucket full at the supermarket, or on Amazon? What ia an amp, or a volt? Eddie Current will help guide you thru these complex subjects.

In the beginning, Nichola Tesla called them electrons before he new they had their own personalities. Now, you have the opportunity to learn all about electrons and will meet their leader, Eddie Current.

Maybe he is a Super-Hero, invisible, but doing all the things that we are dependent on. But where do all these Eddie Current come from??? Where are they stored???

Let me think,,, It seems that in almost every discussion about electricity the word 'Ground' comes up. I know, this only comes up when talking about how to hook up an appliance, or wire a house, but it is a common term. Like you need the black wire hooked up to the 'Hot' and the white wire hooked up to the 'Ground.' Or when a fuse is blown, another term is 'short circuit,' which probably means how long does it take the electrons to get to 'Ground.'

Continuing this train of thought, electrical inspectors regularly check if the installation is properly Grounded.'

This may be a hard concept to visualize. I'm sure you have probably dug in the garden or planted flowers around the house, and in all your digging in the dirt, I'll bet you never uncovered an electron!

Does Eddie Current have the ability to travel invisibly or is HE the *FORCE!*

If you hear someone referring to electricity as "Juice," please demonstrate your expertise in the subject and inform them that Juice is a meaningless colloquial slang term. It should not be used in any serious discussion about electricity.

Electricity has to do with the presence and movement of electrons. Electricity is not power, it is a "carrier" of power.

Let's start with a few explanations,,,

Eddie Current often travels into the ground because it is an easy path and there he finds a home where he is welcomed, loved, given food and a place to recharge.

The major reason that all homes have a 'Ground Wire' is to give Eddie Current a two-lane highway to his favorite restaurant.

"A sweater I bought was picking up static electricity. So, I returned it and They gave me another one free of charge."

The Power Surge:

Lets examine another **Electrical Term**.

A power surge is when Eddie Current gets aggravated by the false theories are trying to take hold. His adrenalin surges and the elevated levels of power will generally cause TVs, smart phones and iPads to turn off, or, in technical terms, cause a non-passive failure.

Now that he has your attention, he then pauses for a few moments to let the climate activists cool off.

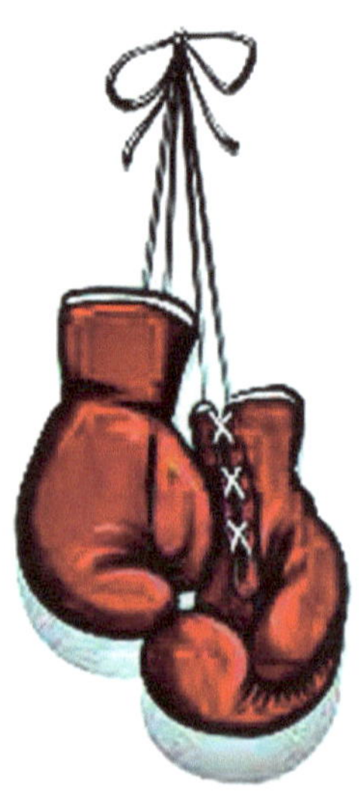

Examples of common household
"Surge Protectors"
consult Sparky, the Electrician
for installation,,,

Electrical Terms: Faults vs Shorts

Shorts are much more common in the summer when the weather is hot and folks wear clothing more appreciate to the temperature. If the temp is around 50 degrees, then nobody is wearing shorts...
Faults could be a blameless crime and regardless of what you are wearing, and you could still be considered clueless...

If your power is out it could be caused by either a short or a fault.

Electric faults could be a blameless crime. Even the geologists who look for faults could be clueless!

Whereas electric shorts are caused by electrons eagerly trying to get back to their home in the ground.
If you are where lots of trees were downed by a hurricane, do you have a short or a fault? You can see the cause, therefore, this is a short.

For the clarification and an actual definition, we asked Eddie Current to weigh-in:

Here is a definition you can take to the bank:
A short circuit has zero resistance between two Wires / Circuits / Systems, on the other hand in a Fault, current has a resistance that draws current. The amount of resistance decides how much current is drawn and could be caused by a breakdown in the insulation of a system...

Are you now ready to explain this to your friends.

Probably more than you wanted to know,
now, whose fault is this???

CLIMATE CHANGE

WE CAN'T FIGHT
HOMELESSNESS, HUNGER, OR POVERTY
BUT WE ARE GOING TO FIGHT "CLIMATE CHANGE"
JUST LET THAT SINK IN FOR A MINUTE.

Another load of electric car fuel headed to the power plant.

What is greenhouse gas?

A day does not go by without you hearing the term 'Climate Change' uttered, usually as the blanket explanation for all of our wows. From natural disasters to man-made catastrophes, the news media to social media to casual conversations all blame 'Climate Change', with the implied and often stated solution of government mandated regulations.

There's a full-blown crisis, mandating conversion to green. The government is mandating, conversion to green without considering the cost, and the dictation of this conversion is eliminating free choice of American citizens. Vehicles emit carbon dioxide, the most common human-caused greenhouse gas. What is greenhouse gas? Surprise! Its mostly carbon dioxide... But most people realize that carbon dioxide is actually a benefit to the planet, so they had to change the name to greenhouse gas.
Its liked a full-blown imitation of Howard Hughes going totally off radar.

We are inundated with statements like: California's just-announced plan to phase out sales of new combustion engine vehicles by 2035. This is doomed to fail. "How can the state electrify the vehicle fleet if it can barely keep the lights on?"

California recently asked homeowners to reduce electricity consumption to help avoid blackouts as temperatures soared and the power system struggled to keep up. The plea was effective, with consumers temporarily dialing back demand enough to keep the lights on across the state. But these sorts of close calls are the stuff of nightmares for system operators, and this specific brush with near-disaster had a new element that caught a lot of attention: a call to electric vehicle owners to avoid charging during peak demand hours.

These types of discussions get emotional quickly, so it's worth stepping back a bit to look at the data.

The world is at a turning point where it must switch from fossil fuels to renewable energy sources in order to tackle climate change. Because they are clean, plentiful, and sustainable, renewable energy sources like wind and solar energy are gaining popularity. Yet, one of the primary issues with renewable energy is the intermittent nature of these sources and these energy storage options.

Solar energy systems can only generate electricity during the day, while wind turbines can only do so when there is enough wind. Because of this, it is challenging to rely on these sources for a regular and reliable energy supply.

Thankfully, technology has improved and a variety of renewable storage options are now becoming more readily available. We shall examine the various forms of renewable energy storage on later pages.

https://www.bloomberg.com/news/articles/2022-09-07/a-text-alert-may-have-saved-california-from-power-blackouts

Please note,,,
This is where we all learn to put on socks
and tie our shoes,,,

We all long for a bright sunny day…

Jonny Nash's 1972 hit, **I Can See Clearly Now**

I can see clearly now the rain is gone
I can see all obstacles in my way
Gone are the dark clouds that had me blind

It's gonna be a bright (bright)
Bright (bright) sunshiny day
It's gonna be a bright (bright)
Bright (bright) sunshiny day

Oh, yes I can make it now the pain is gone
All of the bad feelings have disappeared
Here is that rainbow I've been praying for

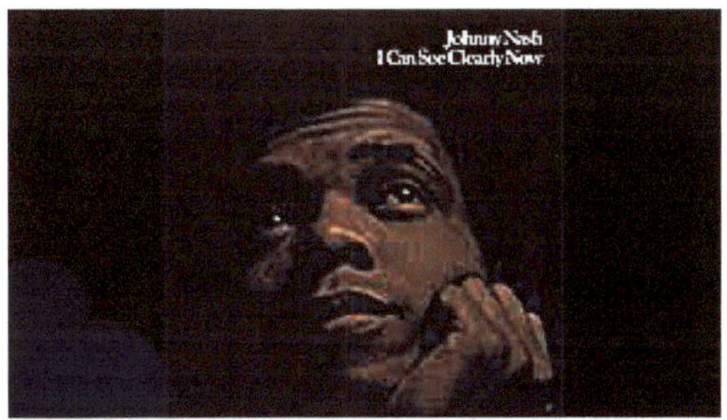

It's gonna be a bright (bright)
Bright (bright) sunshiny day

look all around, there's nothing but blue skies
Look straight ahead, there's nothing but blue skies

I can see clearly now the rain is gone
I can see all obstacles in my way
Here is that rainbow I've been praying for

It's gonna be a bright (bright)
Bright (bright) sunshiny day

Bri-ri-ri-ri', bright, (bright)
Bright (bright) sunshiny day, yeah, eh

let's make some electricity, yeah,, yeah,,, yeah,

Eddie Current is not speaking out against prevailing public opinion. He is not denying that Climate Change is taking place, but it is obvious that electricity is at the center of any discussion of climate change. This is just a small opportunity to give a voice to other causes and alternatives. With a degree of humor, many of the climate change contributors will be exposed and examined, and you will have the opportunity to form you own opinions.

There is actual data that shows that there is climate change. He is only trying to show that the climatologists may be incorrect in that their fundamental assumptions for the causes are too narrow and the proposed solutions may not be founded in logical reason.

Much of their argument is merely a speed bump the highway in attempting to find a solution. When you apply sound scientific and engineering principals to global climate change, you can avoid much of the road kill along the highway to a better environment.

When causes and benefits of corrections are no longer driven by a fundamental assumptions for anthropogenic causes (caused by human activity), solutions will become driven by the desire for an improved global world and will not require government issued incentives. This discussion also validates an assertion that government funding of academic research will also get corrupted by allowing a similar orthodoxy.

Another interesting statement is on the necessity for STEM fields (Science, Technology, Engineering, and Math) to conform to certain dogmas, like anthropogenic climate change. Academics must also change.

Are we living in a POW camp? Or getting our guidance and leadership from a very badly dressed old man?

If you are not a climatologist, prevailing orthodoxy is ostracism or excommunication from an existing climatology society.

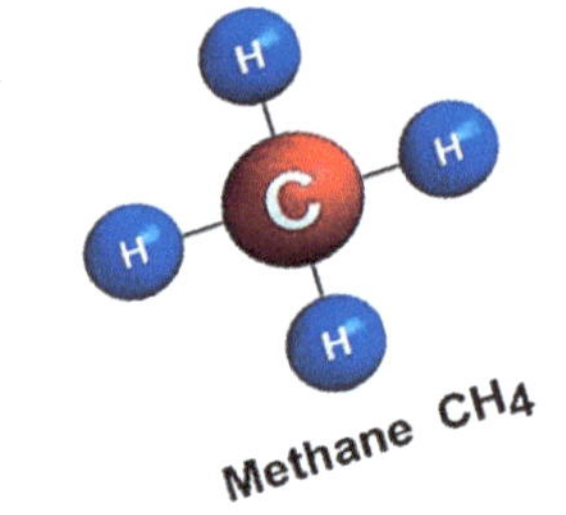

Greenhouse Gases Emitted by Volcanoes, Vehicles, Fossil Fuel Power Plants, Farm Animals, & Rocket Launches

Where does all this CO_2 come from??? A partial list of the culprits include Greenhouse Gases Emitted by Volcanoes, Vehicles, Fossil Fuel Power Plants, Farm Animals, & Rocket Launches.

Which headline will grab your attention? And we all know that the news we are exposed to is driven by advertising dollars and viewer ratings.

Lets start with volcanos. Volcanoes release large amounts of greenhouse gases such as water vapor and carbon dioxide. There have been times during Earth history when intense volcanism has significantly increased the amount of carbon dioxide in the atmosphere and caused global warming.

Today, there are more than 1500 active volcanoes on Earth, with 40-50 continually erupting. Generally around 20 will be aggressively erupting on any particular day .

Once the news media has devoted their maximum 90 seconds to the tragedy, they are on the next flash to keep the viewers attention. We never hear about pollution from farm animals, or jumbo jets!

Thinking back to my formative years, I remember a science class in high school. Our chemistry teacher, Louis Pasture, gave us the experiment for the day. We learned that when mixing baking soda and vinegar, it creates a gas called carbon dioxide. Maybe all the high school science teachers globally are the major culprits in the creation of all this CO_2!

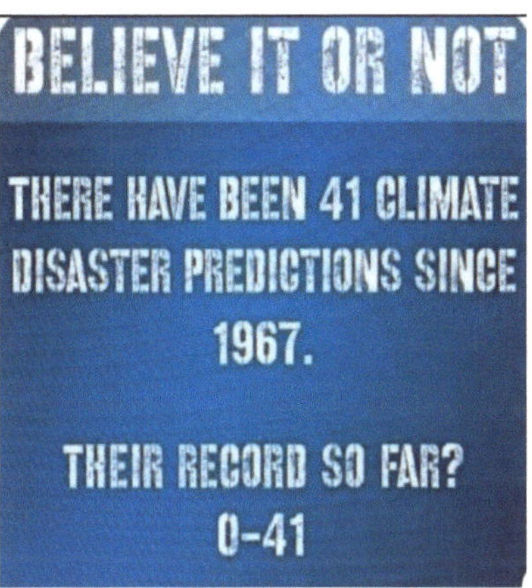

Do you think we should start a fundraiser to send a solar-powered flashlight to every person who still believes the earth is flat.

(I hope they endorse this book)

The most powerful governments on earth can't stop a virus from spreading... but they say they can change the earth's temperature if you pay more taxes?

Causes and Solutions:

Greenhouse Gasses:

While it is tempting to blame the media for over-simplifying complicated scientific ideas and presenting only the bad news, we must remember that they are catering to the desires of their readers and viewers.

Lets take a look at Greenhouse gasses.

What is CO_2

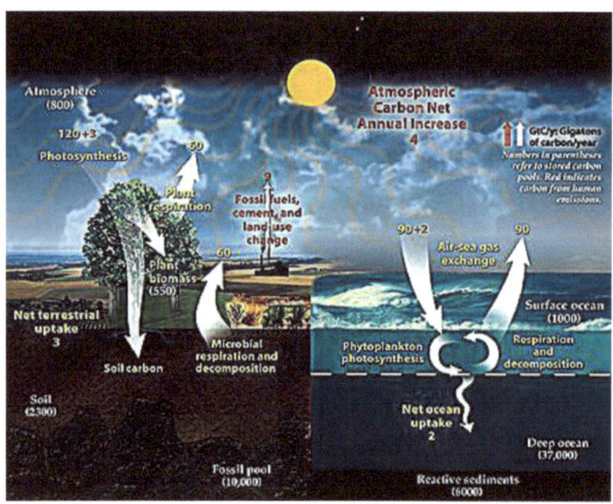

Carbon dioxide is an important greenhouse gas that helps to trap heat in our atmosphere. Without it, our planet would be inhospitably cold. CO_2 is one of the foundations of our world.

For eons, the world's oceans have been sucking carbon dioxide out of the atmosphere and releasing it again. The ocean takes up carbon dioxide through photosynthesis by plant-like organisms phytoplankton), as well as by simple chemistry: carbon dioxide dissolves in water.

Let's explore this narrative about climate change. Common and often used terminology includes the term "Green House gasses." The major gas in this vehicle emission is —- you guessed it—CO_2, a necessity for sustaining life on this planet. Trees convert CO_2 to carbon and oxygen. The oceans need oxygen, fish need oxygen, and with more oxygen, we get more and bigger fish, bigger trees. In fact everything is much better, bigger and healthier.

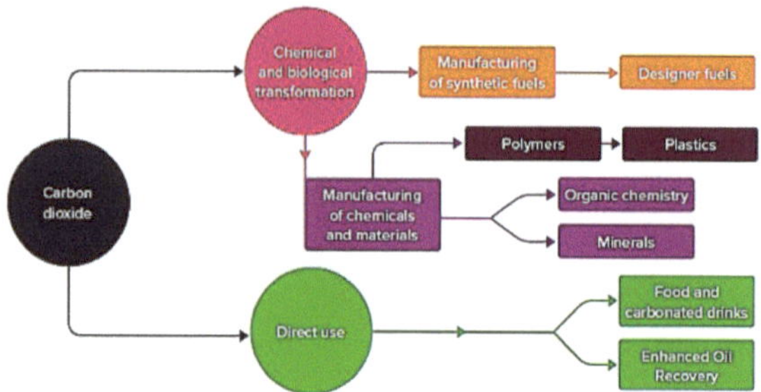

Another "Green House gas is Methane CH_4
Its colorless, odorless and invisible to the naked eye. Methane (a complex hydrocarbon) is another of the foundations of our world, has important implications for climate change, particularly in the near term.

Two key characteristics determine the impact of different greenhouse gases on the climate: the length of time they remain in the atmosphere and their ability to absorb energy. Methane has a shorter atmospheric lifetime than CO_2 , but it is a much more potent greenhouse gas, absorbing much more energy while it exists in the atmosphere, especially the stratosphere..

CH_4 is responsible for more than 25 per cent of the global warming we are experiencing today. Due to its structure, methane traps more heat in the atmosphere per molecule than carbon dioxide (CO_2), making it **80 times more harmful than CO_2** for 20 years after it is released.

Methane is also a great energy source. Remember when we were being told that we would run out of gas, well, now we have found a way to harvest vast amounts of this precious energy source.

Methane comes from many sources. It exists far below the surface and well below the possible existence of any fossils from prehistoric dinosaurs or trees. In fact, it is probably the source of all the shale oil that we are retrieving today.

We haven't been told the truth about where hydrocarbons come from. Climate advocates claim human activity is causing climate change. Other scientists tell us that methane is a far worst gas in the atmosphere, but it is occurring naturally. Methane is found in shale gas.

Shale gas is a form of natural gas (mostly methane), found underground in shale rock. and it is generally distributed over a much larger area. Typically, shale gas consists of 70 to 90 per cent methane (CH_4), the main hydrocarbon target for oil exploration companies. But there is much more in the universe than what is found on Earth. Look no further than Mars.

CH_4 is found with oceans on Titan, a moon of Saturn. So what does that tell us???
Titan, the largest moon of Saturn, is currently a topic of scientific assessment and research. Titan is far colder than Earth, but of all the places in the Solar System, Titan is the only place besides Earth known to have liquids in the form of rivers, lakes, and seas of methane on its surface. Its thick atmosphere is chemically active and rich in carbon compounds. And the temperature is -270^0 F, so it is doubtful that trees and animals lead to this creation.

A review of other scientific literature leads to the conclusion that increases of CO_2 during the twentieth century have produced no deleterious effects upon global climate or temperature.

Increased CO_2 has, however, markedly increased the growth rates of plants as inferred from numerous laboratory and field experiments. There is no clear evidence of the global effects of CO_2 on climate.

Meaningful assessments of the environmental impacts of anthropogenic CO_2 are not yet possible because model estimates of global in climate oncentennial time-scales remain high!

Scientists have been studying our solar system. They have know found that Saturn has more oil and natural gas than Earth and, get this, no evidence of were dinosaurs, plankton and forests on Saturn.

Once again, what do Saturn's moons tell us? Titan has oceans of methane. and why isn't this commonly known and what are the implications of it and what does it tell us about our modern climate change policy.

Willie Soon, **an Astro physicist** for over 30 years, has been thinking about this for a long time, and has authored and published a number of technical articles about this subject.

So why is eliminating CO_2 better for the planet??? Its not, that is probably why the environmentalists couldn't explain methane and had to **change CO_2 to "green house gasses."**

Of particular interest are comments on how the dark ages started,,, when a clerical elite suppressed common discussion and required blind adherence to a narrowly constrained dogma. Does this sound like the media/government now?

So, maybe "green house gasses." are an integral part of our environment, and are not caused by human activities**.**

Causes and Solutions:

Now is the time to take a look at the polluters and the activities that cause these Green House Gases.
China is clearly the largest climate polluter country, making up more than 30% of global emissions.
The top 10 global climate polluters are headed by China and India, and the U S is also in the top 10.

Most solar panels are made of silicon, which is the main component in natural beach sand. Silicon is abundantly available, making it the second most available element on Earth. However, converting sand into high grade silicon comes at a high cost and is an energy intensive process. We never hear of the amount of energy needed to produce a solar panel. The process to produce metallurgical grade requires the same amount of power equivalent to using your home oven for 7 hours.

The total picture— most panels come from China, heat at 1/3 the temperature of the sun, from electricity from coal fired power plants, how much pollution???

Currently, 15% of the world's silicon goes into solar panels.

.

Pollution from the Space Race:

But there's a lot more to rocket pollution than CO_2. Rockets can produce soot (black carbon), nitrogen oxides, alumina particles, chlorine, hydrochloric acid and water vapors. Rocket launches used to be rare enough that pollution wasn't much of a concern.
Today, according to a study by the National Oceanic and Atmospheric Administration (NOAA), **global rocket launches (of which there were 180 last year**, the study notes) inject about 1,000 tons of soot into the upper atmosphere per year. This will only get worse, NOAA warns, as the space race continues to expand.

Many believe that if rocket launches continue to occur at the current rate that they are – it could provoke the temperature of the stratosphere to rise almost three degrees Fahrenheit while still thinning the ozone layer.

The next big headline will probably be about the stratosphere.

Does this mean that "Tesla and Space X Broke the Sky!"

*https://greenly.earth ›
ecology-news
Mar 10, 2023*

https://newsinteractives.cbc.ca

We all have been to the beach, and standard equipment includes SPF30+++. We need some type of SunProtectionFactor for the earth. Maybe CO_2 will work —- OH,, But that is a greenhouse gas and we need to delete it. Sooo, what about a **beach umbrella???** This is certainly a tried and tested solution. More people need to get on board with this idea! But, there is not enough money for them, I guess.

Could a giant space parasol help solve global warming?

Earth is at its hottest point on record, and experts say humans are not doing enough to stop its overheating. Later, we will take a look at the countries causing the most pollution.

But for now, a small but growing number of scientists are proposing a potential solution that could have leaped from the pages of science fiction: The **equivalent of a giant beach umbrella, floating in outer space.**

The idea is to create a huge sunshade between the Earth and the sun to block a small but crucial amount of solar radiation — just enough, the scientists say, to keep Earth within manageable climate boundaries.

Does this solution suggest that
Climate Change is a caused by the SUN???
And it may cost less than $7 Trillion

The New York Times <nytdirect@nytimes.com> 2/3/2024

Is this another undocumented claim of the news media and climate activists?

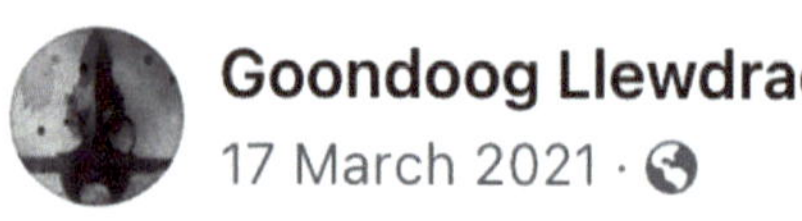

Goondoog Llewdrac
17 March 2021 · 🌐

Climate change crisis? Photo or it didn't happen

Eddie Current (remember, he's been around for a looong time) reminds us that the source if this photo is not verified., and the timing might, in itself, account for the similarity. Timing like high tide and low tide for example,,,

Global Warming

The Arctic Ocean is warming up, icebergs are growing scarcer and in some places the seals are finding the water too hot, according to a report to the Commerce Department yesterday from the Consulate at Bergen, Norway.

Reports from fishermen, seal hunters and explorers all point to a radical change in climate conditions and hitherto unheard of temperatures in the Arctic zone.

Exploration expeditions report that scarcely any ice has been met as far north as 81 degrees 29 minutes.

Soundings to a depth of 3,100 meters showed the gulf stream still very warm

Great masses of ice have been replaced by moraines of earth and stones, the report continued, while at many points well known glaciers have entirely disappeared

Very few seals and no white fish are found in the eastern Arctic, while vast shoals of herring and smelts which have never before ventured so far north, are being encountered in the old seal fishing grounds.

Within a few years it is predicted that due to the ice melt the sea will rise and make most coast cities uninhabitable.

I must apologize.

I neglected to mention that this report was from November 2, 1922 as reported by the AP and published in The Washington Post over 100 years ago.

This must have been caused by Model T Ford emissions or possibly from horse farts and what the buffalo did when they once roamed the Great Plains???

Global Warnings are like pizzas. They come in all shapes and forms: some are small or large circles or squares, thin crust, thick crust, stuffed crust, extra toppings.

Causes and Solutions, continued:

BlockChain Background Information:

A blockchain is a public digital ledger which can be used to verify the ownership of digital assets such as cryptocurrencies.

However, many blockchains, and in particular some cryptocurrencies, use a validation mechanism known as 'proof-of-work'. This mechanism requires vast computing power and results in substantial electricity consumption. For Bitcoin alone, more than 100 TWh per year is now consumed, which is equivalent to the annual electricity consumption of the Netherlands.

As of March 2024, it is difficult to provide an exact number of Bitcoin companies as the industry is constantly evolving and new companies are emerging. However, it is estimated that there are 13,200+ companies working within the blockchain ecosystem, ranging from exchanges and wallet providers to mining operations and blockchain development firms.

Let's explores the idea of who's mining bitcoins?

While it's understandable that you might be curious about the impact that aliens mining bitcoins could have on the market, it's important to remember that this is a highly speculative scenario that is not supported by any evidence. Therefore, it would be fruitless to engage in speculation about how such an event might affect the market.

Did you know that one of the world's first coins had a bee symbol?
Theregoesmymind!

Do you think there's a possibility that Martians are secretly mining bitcoins, given its mysterious value and the vast amount of energy consumption involved?

Digital Mining Energy Costs

The real world cost of digital mining of bitcoin can be quite significant. A major cost is the amount of energy that is required to mine bitcoin. The process involves solving complex mathematical equations, which requires a lot of computational power.

This computational power is provided by specialized computers, which consume a lot of energy.
The energy consumption associated with bitcoin mining is huge. Terawatt-hours per year range are 80+. This equals the energy consumption of an entire countries like Austria or Chile. This has led to concerns about the environmental impact of bitcoin mining, as it contributes to greenhouse gas emissions and the depletion of natural resources.

As demand for bitcoin increases, the price of mining equipment also tends to rise, making it more expensive for miners to acquire the necessary tools and resources.

Do you think there's a possibility that aliens are secretly mining bitcoins, given the vast amount of energy consumption involved.

If bitcoin miners were to switch to pedal power instead of using electricity, how many hours of cycling do you think it would take to mine one bitcoin?

Well, that's an interesting question! While it's difficult to give a precise answer, we can estimate the amount of energy required to mine one bitcoin and then determine how many hours of cycling it would take to generate that same amount of energy.

According to recent estimates, it takes approximately 10,000 to 15,000 kilowatt-hours of electricity to mine one bitcoin. Currently, around 19 million bitcoins have been mined and are in circulation.

Assuming that a person can produce around 100 watts of power while cycling, it would take roughly 100,000 to 150,000 hours of continuous cycling to generate the same amount of energy needed to mine one bitcoin. To put that in perspective, that's the equivalent of cycling non-stop for over 11 years!
This is also the equivalent to the amount of energy used by an average American household in a year!

*Do you think **Cycling for Energy Production** will become an Olympic Sport???*

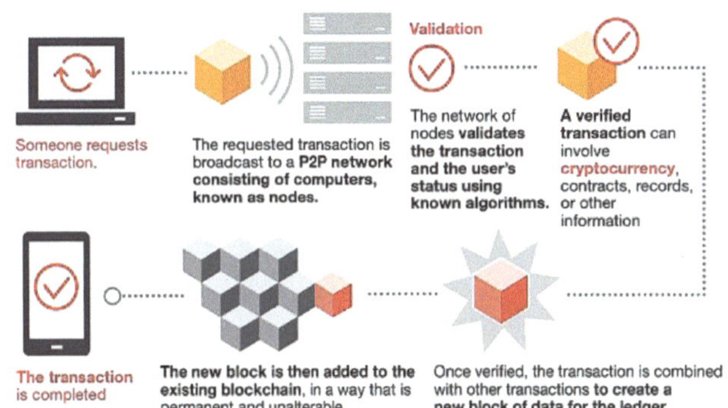

A SIMPLE DIAGRAM OF THE PROCESS

Solar panels and wind farms can generate electricity without releasing any greenhouse gas emissions. Nuclear power plants can too, although today's plants generate long-lasting radioactive waste, which has no permanent storage repository. Please keep in mind, with current alternatives, we do not have enough history to know the recycling costs of lithium batteries and wind turbine blades.

Most nuclear reactors can operate for very long periods of time – over 60 years in many cases.

Here is a nuclear power plant in Ohio that was shut down on 2020.

Although they require a lot of maintenance, some nuclear power stations are now certified for 80+ years of operation, far longer than a gas or coal generator. A fact that is not talked about in the media…

The **problem** is that we are not building any new nuclear plants, *but we can spend $7 trillion on climate change, caused by the power generation plants… Go Figure!*

How is electricity generated using nuclear? - National Grid ESO

Causes and Solutions, continued:

Energy is the centerpiece of most discussions about climate change,,, sooo, Eddie Current is supervising the creation of this document. How is he doing so far?

Recent data shows that electricity usage in the US has increased 28.6%. Another challenge in gaining access all of the renewable energy, the United States needs to dramatically expand the electric grid between places with abundant wind and sunshine and places where people live and work.

More than 20% of the renewable energy installed is **not** connected to the grid. Recent reports show that the U.S. electric grid isn't running 100% of renewable energy yet.

Your guide to the World Economic Forum's annual meeting...

Why is Climate Change the second chapter in a detailed study about electricity? Eddie Current has decided to jump into the fight with his vision of all that is happening.

What is Davos and why is it important?

The WEF, (World Economic Forum), has held a meeting every year since it was founded in 1971. But why is Davos, as it is commonly known, so significant?

Most of the year, Davos is unremarkable other than being a popular ski resort high in the Swiss Alps.

But for one week in January, it becomes the focus of the world's attention as the global elites all converge on the small alpine town for the annual meeting of the World Economic Forum (WEF). Why? To discuss the future direction of life on our planet and the pressing issues of the day.

By David Walsh Published on 15/01/2024

10 years ago, the production of electricity in the US was 82% based on fossil fuels,,, today, 10 years later, 81% of our electric power is produced by fossil fuels -- at an investment of $7,000,000,000, that's $7 trillion dollars on renewable sources!!!

Reported at the Davos summit Jan 2023

"Take advantage, save thousands of dollars on climate-friendly investments through tax credits and rebates."

Reported March 2024:

Global investment in the energy transition hit $1.8 trillion in 2023, up 17% on the previous year and a new record. The total renewable sources investment now exceeds $8.8 trillion!
Energy Transition Investment Trends 2024 | BloombergNEF (bnef.com)

Disaster Capitalism: A Classic Deep Dive,,,

The exploitation of natural or manmade disasters (such as catastrophic weather events such as, war, epidemics, etc.) in service of capitalists interests: a practice of using unstable social, political, and economic situations to impose or benefit from destruction, the privatization of public interests etc.
Slogan: Burn, Flood, or Shake it to the ground!

Let's zoom in on Chile:

Wildfires cause huge loss of life in Chile in South America

 Devastating wildfires have caused dozens of casualties. The disaster occurred as Chile and other parts of South America are gripped by intense heatwaves exacerbated by climate change.

Then add:
Chilean President Gabriel Boric announced preliminary plans for a national lithium strategy, one that would increase government participation in the lucrative sector. Chile is currently the world's second-largest producer of lithium.

Then add:
Currently, only two companies mine lithium in Chile: Chilean firm SQM, which is nearly a quarter-owned by a Chinese company, and U.S. firm Albemarle. They are both private and exclusively operate the country's more than 40 salt flats that are home to lithium reserves.

(*Detailed reference data, like this, can be found throughout*)
https://m.youtube.com/watch?v=pr0LkPMZ-qc
https://youtu.be/QsiGiziM7Vw?si=7ubshkj3YmQHXpRo

(*foot notes from Eddie Current*)
As Abe Lincoln once said, 'If it is on the internet,,,
then it must be true.'

An early **Climate Investigator** and would-be aircraft designer/inventor took the initiative seeking fame and glory.

Do you think he was trying to measure Methane in the stratosphere???

Lawn chair technology's early success stories start with Larry Walters. Forty-two helium-filled weather balloons lifted him in this aluminum lawn chair from San Pedro, California, on July 2, 1982.

Lawn chair balloon pilot. **Lawrence Richard Walters**, nicknamed "Lawn Chair Larry" or the "Lawn Chair Pilot", (April 19, 1949 – October 6, 1993) was an American truck driver. On 2 July 1982 he attempted to reach outer space in a homemade aircraft.

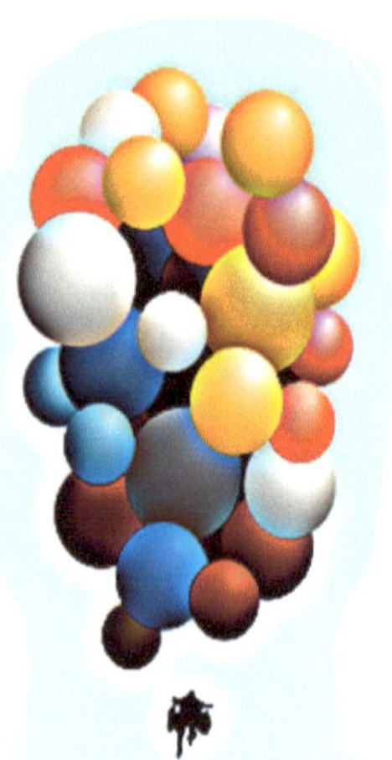

Larry Walters, a.k.a. Lawnchair Larry, rode a lawn chair attached to 42 helium-filled weather balloons in 1982, soaring up to about 4.9 km.

Facing freezing temperatures and lower oxygen levels, Walters popped balloons with a BB gun to establish a controlled descent. The Federal Aviation Administration charged him with violating controlled air- space, flying without a balloon license, and operating a non-airworthy craft.

Eddie Current's version would most likely state 'A non-Worthy Government foolishly charged a highly creative Citizen with an over abundance of Initiative!'

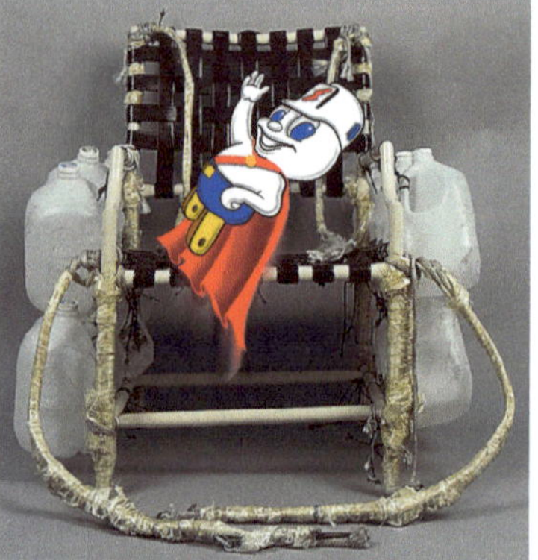

National Air and Space Museum

https://airandspace.si.edu
Chair, Lawn, Larry Walters

Methane produced by "cow burps" is one of the biggest sources of greenhouse gas emissions from agriculture. A single cow produces between 154 to 264 pounds of methane gas per year. Studies have shown that grass-fed cattle produce 20% more methane in their lifetime than grain-fed cattle. Not counting for the emissions of any other livestock, 1.5 billion cattle, raised specifically for meat production worldwide, emit at least 231 billion pounds of methane into the atmosphere each year.

And what about the English & Irish with all their sheep??? https://www.epa.gov › snep › agric…

2024, Timely notes & Last minute updates:

#1. When talking about the growth of renewable energy, I often say it's not notable when wind, solar or other technologies reach a record high, because they should be doing that every year. Then 2023 came along. For the first time in more than a decade, the United States had a **decrease in utility-scale electricity generation** from renewable sources, according to the Energy Information Administration.

#2. Puget Sound Energy has filed with the Public Utilities Commission for a **20% rate** increase to cover $530 million the **costs of wind turbines**.

#3. NY attorney general sues world's largest **beef producer** over methane emissions, climate commitments.
NY AG says company's net zero goals aren't feasible considering its high carbon footprint

#4. THE WORLD'S LONGEST ONSHORE **WIND TURBINE BLADE** CREATES A DIAMETER AS LONG AS **THREE FOOTBALL FIELDS** -AND IT'S SET TO DEBUT SOON (That is more than 900ft) The blade, which will be attached to a 15-megawatt wind turbine, is super light and super strong.

#5. Global investment in the energy transition hit **$1.8 trillion in 2023**, up 17% on the previous year and a new record. The total renewable sources investment now exceeds $8.8 trillion!
Energy Transition Investment Trends 2024 | BloombergNEF (bnef.com)

#6. In Georgia, **demand for industrial power** is surging to record highs, with the projection of new electricity use for the next decade now **17 times** what it was only recently. Internet, March 2024

#7. Northern Virginia needs the equivalent of **several large nuclear power plants** to serve all the new data centers planned and under construction. March 2024

#8. Only a measly **1% of US retailers** offer EV charging stations even though they come with significant benefits and government incentives, according to Consumer Reports.

#9. The **Chevrolet Bolt** was the little electric car that could. Never the fastest or the fanciest EV, the Bolt and its sticker price of roughly $30,000 made it cheaper than lots of gas cars, all while delivering a respectable 259 miles of range. With its legion of fans, the Bolt outsold every non-Tesla EV in America last year. But if you want to buy a new one, you'd better hurry: Chevy's parent company, General Motors, stopped making the Bolt at the end of 2023.
Electric Cars Are Still Not Good Enough - The Atlantic

#10. The CEO of **Rivian Automotive** announced Thursday that the electric truck maker is pausing construction of its $5 billion manufacturing plant in Georgia to speed production and save money.
Tesla-rival Rivian is pausing construction on its $5bn electric truck plant in Georgia to speed production and save cash | Fortune

#11. Our Electric Grids are Blocking a Climate Revolution
Time.com/6900079/electric-grids-climate-revolution/

#12. Should excessive electricity bills be reported as **a way to identify illegal marijuana** grows in Maine,,, The idea to report high electricity usage comes after frustration from Maine law enforcement who said they can't keep up with the estimated 270 illegal grow operations that have popped up across the state...

POWER COMPANIES

The following chapter is more than scrapes and bits of information, but, it is enough information for you to change a light bulb and pay your power bill.

*Published by Statista Research Department,
Dec 18, 2023*

The day starts with breakfast, usually including toast. Where does Eddie Current go when he leaves the toaster. Lets start this chapter with your power company.

An electric utility is most often a company that operates facilities to generate, transmit, and/or distribute electricity to both private, public and industrial consumers. Electric utilities can be involved in one or more of these activities. Some utilities only function in several categories. TVA, for example, only generates and distributes electricity. Others, like electricity marketers, only buy and sell electricity, can also be considered utilities.

In the United States, there are around 3,**000** electric utility companies providing power to more than **140 million customers.**

Too much detail???

More details are coming,,,

Found on the internet: The world is at a turning point where it must switch from fossil fuels to renewable energy sources in order to tackle climate change. Because they are clean, plentiful, and sustainable, renewable energy sources like wind and solar energy are gaining popularity. Yet, one of the primary issues with renewable energy is the intermittent nature of these sources and these energy storage.

Solar energy systems can only generate electricity during the day, while wind turbines can only do so when there is enough wind. Because of this, it is challenging to rely on these sources for a regular and reliable energy supply.

Thankfully, technology has improved and a variety of renewable storage options are now readily available. We shall examine the various forms of renewable energy storage on the following pages.

https://www.iea.org/commentaries/tripling-renewabke-power-capacity-by-2030-is vital-to keep-the -150c-goal-within-reach

"We need to eliminate global emissions of greenhouse gases by 2050,"
philanthropist and technologist Bill Gates
wrote in his 2023 annual letter...

WOW, 27 years and $7 trillion,,,
feel the sense of urgency yet???

Lets Take a Look at Your Local Power Company:

I hope that the following pages will help shed some light on the complex subject of your power bill, but, if you want to skip these pages, please feel free.

As we look at the charges for the work that Eddie Current does for us across the country, we can see a great variety in charges and rational for providing this service. Some localities have simple, easy-to-understand billing formats, while others desire to complicate the entire process—probably so that they can get away with charging more.

We have selected several states that are typical for their practices and geography. Using the universal measure of quantity, the Kilowatt Hour (KWH), we have demonstrated a range of charges of $0.136/KWH to $0.359/KWH (13.6 ¢/KWH to 35.9¢/KWH).

 California has unusually high electricity rates, additionally, because of state mandates for sustainable energy, mostly hydro-electric from Grand Coulee Dam and the Pacific Northwest. There has been serious underinvestment in conventional nuclear and fossil fuel electricity in favor of solar and wind subsidies.   It is the most expensive electricity in the country.

Lets remind ourselves that Eddie Current is the same guy in every state. The work he does is the same powering lighting, computers, cooking, and TVs. He comes from the same sources of fossil fuels, solar, and wind, but your charges vary greatly.

I hope that the following pages help shed some light on this complex subject.

As we look at the charges for the work that Eddie Current does for us across the country, we can see a great variety in charges and rational for providing this service. Some localities charge have simple, easy-to-understand billing formats, while others desire to complicate the entire

process—probably so that they can get away with charging more.

Here are a few of the highlights of the complex details, Included are:
 ~Fuel and Non-fuel charges,
 ~Delivery charges and Transition charges *(whatever those are)*,
 ~Merchant function charges *(why)*,
 ~Peak and Off-Peak charges *(I think I can understand this one)*,
 ~Court resolution surcharge (*?*),
 ~Meter reading *(that's the basic cost of doing business!)*,
 ~Regulatory assessment (*?*),
 ~Franchise fee *(give me a break—this is not a McDonalds)*

Please remember, the universal measure of quantity is the Kilowatt Hour (KWH), (which we have never seen, let alone even measure) and we have demonstrated a range of 13.6 ¢ to 35.9 ¢.

#1. This **New York State Electric & Gas** power bill from upstate New York is quite confusing,,, I had to read it several times... I believe that the rate in this area averages out to $0.178/KWH...(17.8 ¢/KWH)

And it seems that a lot of government agencies get a piece of the action.,., including, but not limited to, 'transition charges,' 'merchant function,' and 'school dist. sales tax,,,' *whatever these charges mean...*

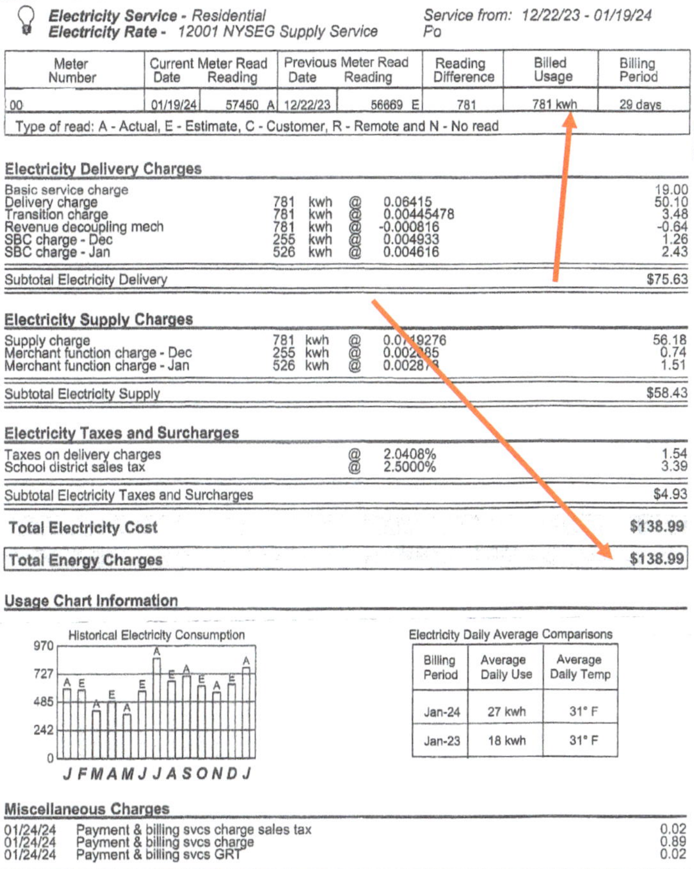

Electricity Service - Residential   Service from: 12/22/23 - 01/19/24
Electricity Rate - 12001 NYSEG Supply Service   Po

Meter Number	Current Meter Read Date	Reading	Previous Meter Read Date	Reading	Reading Difference	Billed Usage	Billing Period
00	01/19/24	57450 A	12/22/23	56669 E	781	781 kwh	29 days

Type of read: A - Actual, E - Estimate, C - Customer, R - Remote and N - No read

Electricity Delivery Charges

Basic service charge				19.00
Delivery charge	781 kwh	@	0.06415	50.10
Transition charge	781 kwh	@	0.00445478	3.48
Revenue decoupling mech	781 kwh	@	-0.000816	-0.64
SBC charge - Dec	255 kwh	@	0.004933	1.26
SBC charge - Jan	526 kwh	@	0.004616	2.43
Subtotal Electricity Delivery				$75.63

Electricity Supply Charges

Supply charge	781 kwh	@	0.0719276	56.18
Merchant function charge - Dec	255 kwh	@	0.002885	0.74
Merchant function charge - Jan	526 kwh	@	0.002878	1.51
Subtotal Electricity Supply				$58.43

Electricity Taxes and Surcharges

Taxes on delivery charges	@	2.0408%	1.54
School district sales tax	@	2.5000%	3.39
Subtotal Electricity Taxes and Surcharges			$4.93

Total Electricity Cost	$138.99
Total Energy Charges	$138.99

Usage Chart Information

Historical Electricity Consumption

Electricity Daily Average Comparisons

Billing Period	Average Daily Use	Average Daily Temp
Jan-24	27 kwh	31° F
Jan-23	18 kwh	31° F

Miscellaneous Charges

01/24/24	Payment & billing svcs charge sales tax	0.02
01/24/24	Payment & billing svcs charge	0.89
01/24/24	Payment & billing svcs GRT	0.02
Total Miscellaneous Charges		$0.93

Customer Name: Account Number: FPL.com Page 2

BILL DETAILS

Amount of your last bill	196.24
Payment received - Thank you	-196.24
Balance before new charges	$0.00

New Charges
Rate: RS-1 RESIDENTIAL SERVICE

Base charge:		$9.48
Non-fuel:	(First 1000 kWh at $0.087920) (Over 1000 kWh at $0.097840)	$88.69
Fuel:	(First 1000 kWh at $0.034620) (Over 1000 kWh at $0.044620)	$34.98
Electric service amount		133.15
Gross receipts tax (State tax)		3.42
Taxes and charges		3.42
Regulatory fee (State fee)		0.10
Actual electric charges		136.67
Budget billing charges		
Total amount you owe		

FPL automatic bill pay - DO NOT PAY
EDI File Transmitted Separately

METER SUMMARY

Meter reading - Meter ACD1506. Next meter reading Feb 21, 2024.

Usage Type	Current	-	Previous	=	Usage
kWh used	76970		75962		1008

ENERGY USAGE COMPARISON

	This Month	Last Month	Last Year
Service to	Jan 22, 2024	Dec 20, 2023	Jan 21, 2023
kWh Used	1008	1110	1095
Service days	33	30	32
kWh/day	31	37	34
Amount	$136.67	$152.12	$138.33

FPL BUDGET BILLING

Deferred Balance $54.52

KEEP IN MIND

- Taxes, fees, and charges on your bill are determined and required by your local and state government to be used at their discretion.
- The fuel charge represents the cost of fuel used to generate electricity. It is a direct pass-through to customers. FPL does not profit from fuel, although higher costs do result in higher state and local taxes and fees.

#2. Florida Power & Light...

Here's a relatively easy electric bill to understand. Only 3 line items, Base, Non-fuel, and Fuel.

I believe that the rate in this area averages out to $0.136/KWH...(13.6 ¢/KWH)

ENERGY STATEMENT
www.pge.com/MyEnergy

Account No: ▓▓▓▓▓▓▓▓
Statement Date: 01/10/2024
Due Date: **01/31/2024**

Service For:
▓▓▓▓▓▓▓▓

Questions about your bill?
Mon-Fri 7 a.m.-7 p.m.
Saturday 8 a.m.-5 p.m.
Phone: 1-800-743-5000
www.pge.com/MyEnergy

Ways To Pay
www.pge.com/waystopay

Your Account Summary

Amount Due on Previous Statement	$292.12
Payment(s) Received Since Last Statement	-292.12
Previous Unpaid Balance	$0.00
Current PG&E Electric Delivery Charges	$160.50
Central Coast Community Energy Electric Generation Charges	64.93

Total Amount Due by 01/31/2024   **$225.43**

Electric Monthly Billing History

Daily Usage Comparison

1 Year Ago	Last Period	Current Period
23.07	27.49	20.93

Electric kWh / Day

$300
$225
$150

2/09 3/13 4/11 5/11 6/10 7/12 8/10 9/11 10/11 11/09 12/11 1/10 **2024**

www.pge.com/MyEnergy *for a detailed bill comparison*

ENERGY STATEMENT
www.pge.com/MyEnergy

Details of Central Coast Community Energy Electric Generation Charges
12/11/2023 - 01/09/2024 (30 billing days)
Service For: ▓▓▓▓▓▓▓▓
Service Agreement ID: ▓▓▓▓▓▓▓▓
Rate Schedule: MBRETCH1 3Cchoice ETOUB

12/11/2023 – 01/09/2024

Generation - Peak - Winter	64.553100	kWh	@ $0.15700	$10.13
Generation - Peak - Winter	40.659200	kWh	@ $0.18300	7.44
Generation - Off Peak - Winter	326.240900	kWh	@ $0.08500	27.73
Generation - Off Peak - Winter	196.341900	kWh	@ $0.09900	19.44
Energy Commission Tax				0.19

Total Central Coast Community Energy Electric Generation Charges   **$64.93**

Account No: ▓▓▓▓▓▓▓▓
Statement Date: 01/10/2024
Due Date: **01/31/2024**

Your Electric Charges Breakdown (from page 2)

Transmission	$32.35
Distribution	102.74
Electric Public Purpose Programs	16.53
Nuclear Decommissioning	-0.09
Wildfire Fund Charge	3.40
Recovery Bond Charge	3.31
Recovery Bond Credit	-3.31
Wildfire Hardening Charge	1.59
Competition Transition Charges (CTC)	0.36
Energy Cost Recovery Amount	-0.29
PCIA	3.19
Taxes and Other	0.72
Total Electric Charges	**$160.50**

Details of PG&E Electric Delivery Charges
12/11/2023 - 01/09/2024 (30 billing days)
Service For: ▓▓▓▓▓▓▓▓
Service Agreement ID: ▓▓▓▓▓▓▓▓
Rate Schedule: ETOUB B Residential Time-of-Use Service

12/11/2023 – 12/31/2023

Energy Charges				
Peak	64.553100	kWh	@ $0.39763	$25.67
Off Peak	326.240900	kWh	@ $0.35883	117.07
Generation Credit				-52.58
Power Charge Indifference Adjustment				1.03
Franchise Fee Surcharge				0.43

01/01/2024 – 01/09/2024

Energy Charges				
Peak	40.659200	kWh	@ $0.46592	$18.94
Off Peak	196.341900	kWh	@ $0.42712	83.86
Generation Credit				-36.37
Power Charge Indifference Adjustment				2.16
Franchise Fee Surcharge				0.29

Total PG&E Electric Delivery Charges   **$160.50**

2018 Vintaged Power Charge Indifference Adjustment

Account No: ▓▓▓▓▓▓▓▓
Statement Date: 01/10/2024
Due Date: **01/31/2024**

Service Information

Meter #	▓▓▓▓▓▓▓▓
Total Usage	627.795100 kWh
Heat Source	B - Not Electric
Serial	Q
Rotating Outage Block	50

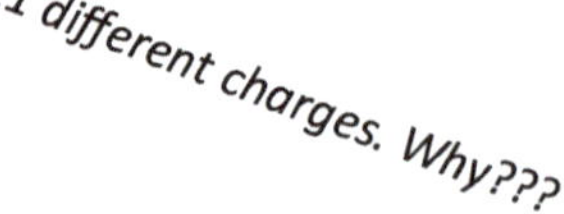

4. Arizona Public Service,,,

I believe that the rate in this area averages out to $0.172/KWH…(17.2¢/KWH),,, and here's how the bill breaks down: **18 line** items for Eddie Current, **5 line** items for governmental getting their fingers into this activity, **4 line** items for time-of-day usage.

It is a conspiracy to undermine and attack our level of understanding.

I hope that these pages help shed some light on this complex subject. As we look at the charges for the work that Eddie Current does for us across the country, we can see a great variety in charges and rational for providing this service. Some localities charge have simple, easy-to-understand billing formats, while others desire to complicate the entire process—probably so that they can get away with charging more. The universal measure of quantity is the Kilowatt Hour (KWH), and we have demonstrated a range of $0.136/KWH to $0.359/KWH.

And increases are in the pipeline, so to speak!

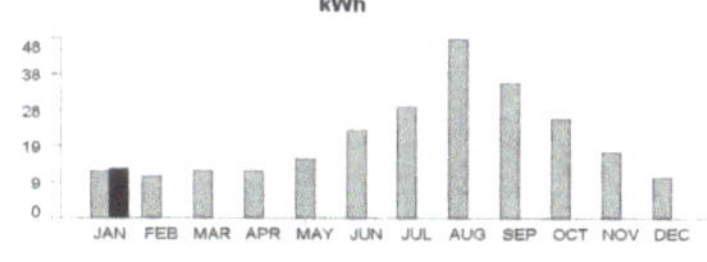

Your electricity bill	Morel	Your account number
January 9, 2024		

Service plan: Time-of-Use 4pm-7pm Weekdays

Meter number:
Meter reading cycle: 06

Charges for electricity services

Cost of electricity you used

Customer account charge	$2.18
Super Off-peak delivery service charge	$0.91
On-peak delivery service charge	$1.23
Off-peak delivery service charge	$8.84
Environmental benefits surcharge	$3.57
Federal environmental improvement surcharge	$0.06
System benefits charge	$1.33
Power supply adjustment*	$8.07
Metering*	$6.02
Meter reading*	$2.18
Billing*	$2.43
Generation of electricity super off-peak*	$0.60
Generation of electricity on-peak*	$9.75
Generation of electricity off-peak*	$18.91
Federal transmission and ancillary services*	$4.64
Federal transmission cost adjustment*	$0.16
Court resolution surcharge	$0.74
LFCR adjustor	$1.11
Cost of electricity you used	$72.73

Taxes and fees

Regulatory assessment	$0.18
State sales tax	$4.16
County sales tax	$0.52
City sales tax	$0.74
Franchise fee	$1.46
Cost of electricity with taxes and fees	$79.79
Total charges for electricity services	**$79.79**

** These services are currently provided by APS but may be provided by a competitive supplier.*

Amount of electricity you used

Meter reading on Jan 9	20426
Meter reading on Dec 8	20003
Total electricity you used, in kWh	**423**
On-peak meter reading on Jan 9	3586
On-peak meter reading on Dec 8	3545
On-peak electricity you used, in kWh	**41**
(4pm - 7pm Monday - Friday)	
Off-peak electricity you used, in kWh	**296**
(All other hours and certain holidays)	
Super off-peak meter reading on Jan 9	4039
Super off-peak meter reading on Dec 8	3953
Super off-peak electricity you used, in kWh	**86**
(10 am to 3 pm Monday - Friday, November - April)	

Average daily electricity use per month

Comparing your monthly use

	This month	Last month	This month last year
Billing days	32	30	33
Average outdoor temperature	54°	63°	51°
Your total use in kWh	423	334	412
Your average daily cost	$2.49	$2.28	$2.22

Your Monthly Plan Comparison	View more details or switch plans by calling (800) 253-9405 or visiting **aps.com/compare**.		
Your Current Plan **Time-of-Use 4pm-7pm Weekdays**	Save The Most With **Fixed Energy Charge Plan - Tier 2**	This Month **$6.29 more**	12 Month Savings **$61.22**

What we need is more agencies involved in this power industry.

The electrical power industry has evolved significantly over the years, with the emergence of various market participants and channels of distribution.

COMPLEXITY? YEP!!!

Eddie Current first was recruited by the TVA (Tennessee Valley Authority) as a Marketing Manager, what started out as a unified industry where the electrons were produced and then distributed to Municipal Power Companies and REAs (Rural Electric Authorities). The electrons went from the producer, were transmitted directly to the distributor, who then delivered directly to our homes and businesses… A simple 3 step supply chain.

Then Government had to get its sticky fingers into the pot. The next step in this process was adding the step referred to as 'Transmission.' this was necessary because the 'factory' was located so far from where we lived and worked.

So, initially, there were three elements in the electron process— 'Generation,'
        'Transmission,'
        'Distribution.'

Federal and local governments quickly saw an opportunity to get a piece of the action and then created Public Utility Commissions (PUC).

Firstly, it is important to note that the electrical power industry has undergone significant changes since the early 1990's. The deregulation has led to the emergence of various market participants.

Get the picture now???
Maybe it is like
wrestling a greased pig!

Here's what the industry looks like today…

1. **Generators:** Where electrons are produced - coal-fired power plants, gas-fired power plants, nuclear power plants, hydro plants, wind farms, and solar farms.

2. **Transmission System Operators (TSOs):** Organizations that operate the high-voltage transmission network giving Eddie Current a path to your neighborhood and can also be referred to as independent system operators **(ISOs)** and regional transmission organizations **(RTOs)** manage grid operations

3. **Distribution System Operators (DSOs):** Organizations that delivers electrons to your door. Also known as Grid Operators , there ae actually 10 covering the USA

4. **Public Utility Commissions (PUC):** Government agencies that give all players a chance to make HUGE profits.

5. **Retailers or Suppliers:** Organizations that buy electricity from the wholesale market and sell it to those who need it.

6. **Aggregators:** These are entities that bundle supply from multiple sources and sell it to end-consumers.

7. **Brokers:** Intermediaries between buyers and sellers.

8. **Market Operators:** Entities that providing a trading platform.

And we find upon closer examination that each of these segments has more than a few sub-categories!

Here is more information than you probably thought you would get with this book, and don't feel obligated to read all this detail.

What is the Grid?

The Grid links many states to insure a more reliable system of supply. Texas is a bit different,,, and ERCOT was formed. What Is ERCOT???

The North American electric power system is configured with 3 grids,,, The Eastern Interconnection, The Western Interconnection, and the Texas Interconnection, the Electric Reliability Council of Texas (ERCOT) is independent.

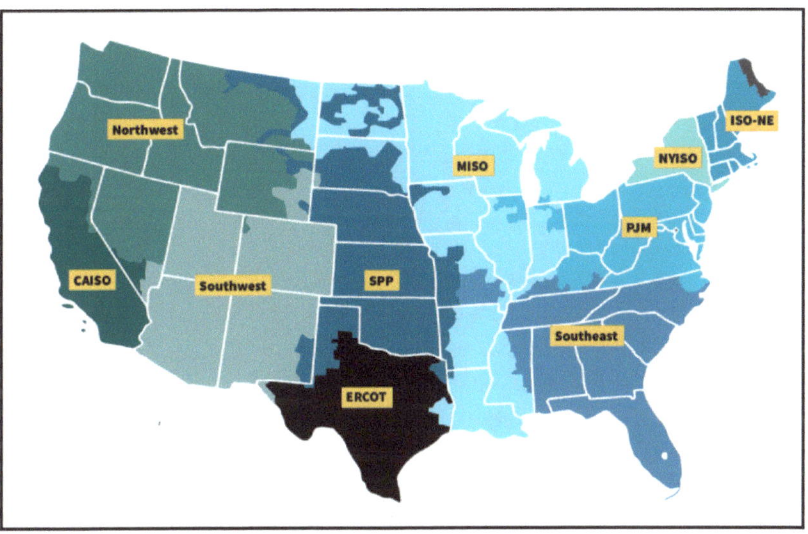

terconnection, and Electric Reliability Council of Texas (ERCOT). Source: FERC (map retrieved Dec. 29, 2023)

—**Aaron Larson** *is POWER's executive editor (*@POWERmagazine*).*

In many parts of the country, independent system operators **(ISOs)** and regional transmission organizations **(RTOs)** manage grid operations. **ISOs** grew out of Federal Energy Regulatory Commission **(FERC),** issued on April 24, 1996. In the orders, the commission suggested the concept of an **ISO** as one way for existing tight power pools to satisfy the requirement of providing non-discriminatory access to transmission. Subsequently, FERC Order No. 2000, issued on Dec. 20, 1999, the voluntary formation of RTOs to administer the transmission grid on a regional basis throughout North America, including Canada.

While major sections of the country operate under more traditional market structures, two-thirds of the nation's electricity load is served in RTO regions. Among the system operators are California ISO (CAISO), Mid-continent ISO (MISO), ISO-New England (ISO-NE), New York ISO (NYISO), Southwest Power Pool (SPP), PJM In-

Number of electricity providers in the United States in 2022, by ownership type

Characteristic	Number of utilities
Investor owned	**166**
Political subdivision	83
Wholesale power marketer	37
Community choice aggregator	26

*What is your **EGO** factor now,,, 10???*
*(**E**yes **G**lazed **O**ver)*

How many coal-fired power plants are there in the world today?

Great question. These are the plants world wide that are supposedly causing the most climate change and are driving the race to renewable generation.

The EU has 468 - building 27 more... Total of 495
Turkey has 56 - building 93 more... Total 149
South Africa has 79 - building 24 more... Total 103
India has 589 - building 446 more... Total 1035
The Philippines has 19 - building 60 more... Total 79
South Korea has 58 - building 26 more... Total of 84
Japan has 90 - building 45 more... Total 135
China has 2,363 - building 1,171 more... Total = 3,534

That's 5,615 projected coal-powered plants in just 8 countries.

USA has 15 - building 0 more...Total = 15

However, in the meantime, China still consumes nearly five times as much coal as India, and nearly six times as much as the United States and is building a huge number of new plants.

China is Currently Building Over Half of The World's New Coal-based Power Plants.

In 2021, China began building 33 gigawatts of coal-based power generation, according to the Helsinki-based Center for Research on Energy and Clean Air (CREA). That is the most new coal-fired power capacity China has undertaken since 2016 and, says CREA, three times more than the rest of the world combined. India, Brazil and Russia are right behind China.

China Has Built 14 Overseas Coal Plants Since Vowing No New Ones,,,

The gears of Climate Change grind slowly, and are driven by electrons.

https://www.bloomberg.com/news/articles/2022-09-22/china-has-built-14-overseas-coal-plants-since-vowing-no-new-ones

https://e360.yale.edu/features/despite-pledges-to-cut-emissions-china-goes-on-a-coal-spree

Chinese Coal-based Power Plants

Industrialization was encouraged by the ready availability of plentiful and abundant cheap energy/ electricity here in the US Here is an example how China is funding its growth in industrializing. Smoking chimneys and solar panel farm are seen along a highway in a coal producing region in Yulin in north-western China's Shaanxi province, April 24, 2023.

Europe has its own unique energy problems. After years of misguided energy policies, Europe's electricity has become so expensive that trade unions have started warning of the threat of deindustrialization. Many of its nuclear plants are being shuttered and new sanctions are disrupting the supply of Russian natural gas. Additionally, the European Union has doubled down on renewable energy mandates, further con-stricting the supply of fossil fuel power.

Europe's electricity prices have settled at triple their pre-pandemic levels, and are projected to remain at this level for some time.

Will this be the unforeseen consequence here in the US???

High electricity prices have Europe facing deindustrialization; don't let it happen here | The Hill OPINION-ENERGY AND ENVIRON-MENT—Jan 2024

The trend so far looks very different in countries with **rapidly growing economies** — nowhere more so than in China.

China overtook the United States as the world's single largest power producer in 2010, and now makes nearly a third of the world's electricity. (The country's per person electricity generation is still much lower than America's.)

(Where did all these electrons come from??? Did Eddie Current's family suddenly grow? Where are all these electrons buried in the ground?)

For decades, the country's soaring power demand was fulfilled largely by coal, the most polluting fossil fuel. Coal-fired generation continued to grow, though at a slower pace, even as China *significantly expanded carbon-free power* in recent years, especially wind and solar.

Energy analysts expect carbon-free power to grow enough in the next few years to start pushing out coal-fired electricity here, too. And because of the country's outsize share of the global total, peak coal power in China will likely be a global one.

How quickly coal power and related emissions will decline after that peak is less clear. In a recent agreement with the United States, China said it would speed up renewable energy deployment to "accelerate the substitution for coal, oil and gas generation" and agreed to pursue "meaningful" power-sector emissions cuts over the next decade. But whether it will stop approving new coal plants remains to be seen.

Electricity Around the World

By Nadja Popovich Nov. 20, 2023 New York Times

Carbon-free electricity has never been more plentiful. Wind and solar power have taken off over the past two decades, faster than experts ever expected. But it hasn't yet been enough to halt the rise of coal- and gas-burning generation.

That's because global demand for electricity has grown even faster than clean energy, leaving fossil fuels to fill the gap.

The dynamic has pushed up carbon emissions from the power sector at a time when scientists say they need to be falling — and fast — to avoid dangerous levels of global warming.

A whole bunch of folks just aren't listening or maybe they just don't care!!!

Do you think that there are countries where electrical connections are really scary???

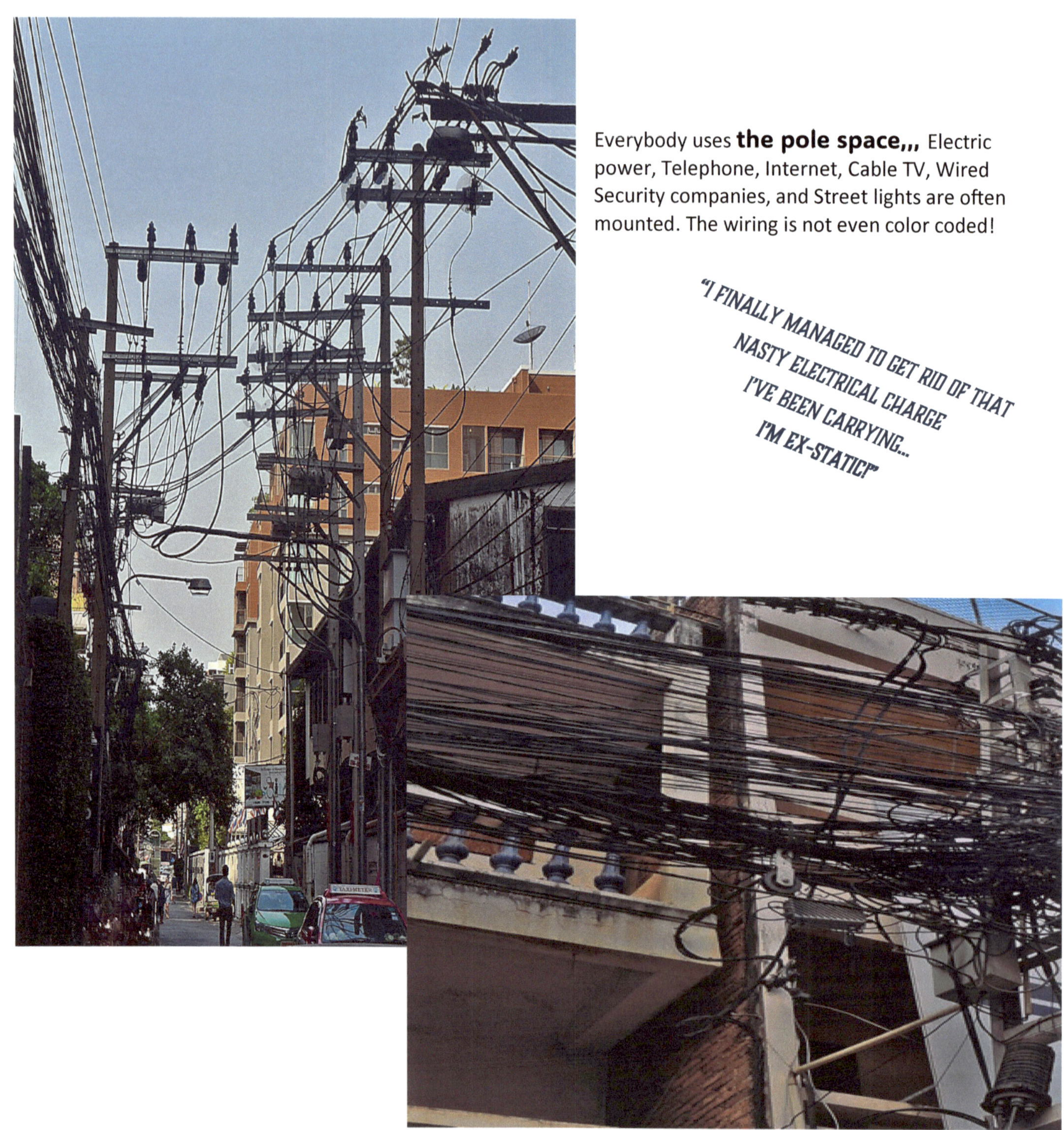

Everybody uses **the pole space,,,** Electric power, Telephone, Internet, Cable TV, Wired Security companies, and Street lights are often mounted. The wiring is not even color coded!

Do you think the local infrastructure needs upgrading???

Do you think this was planned, or did it just happen

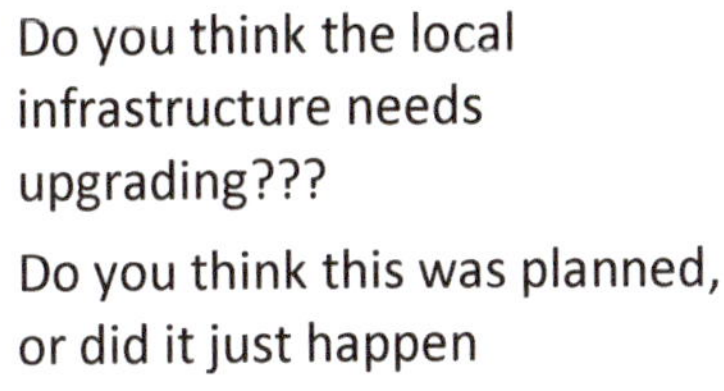

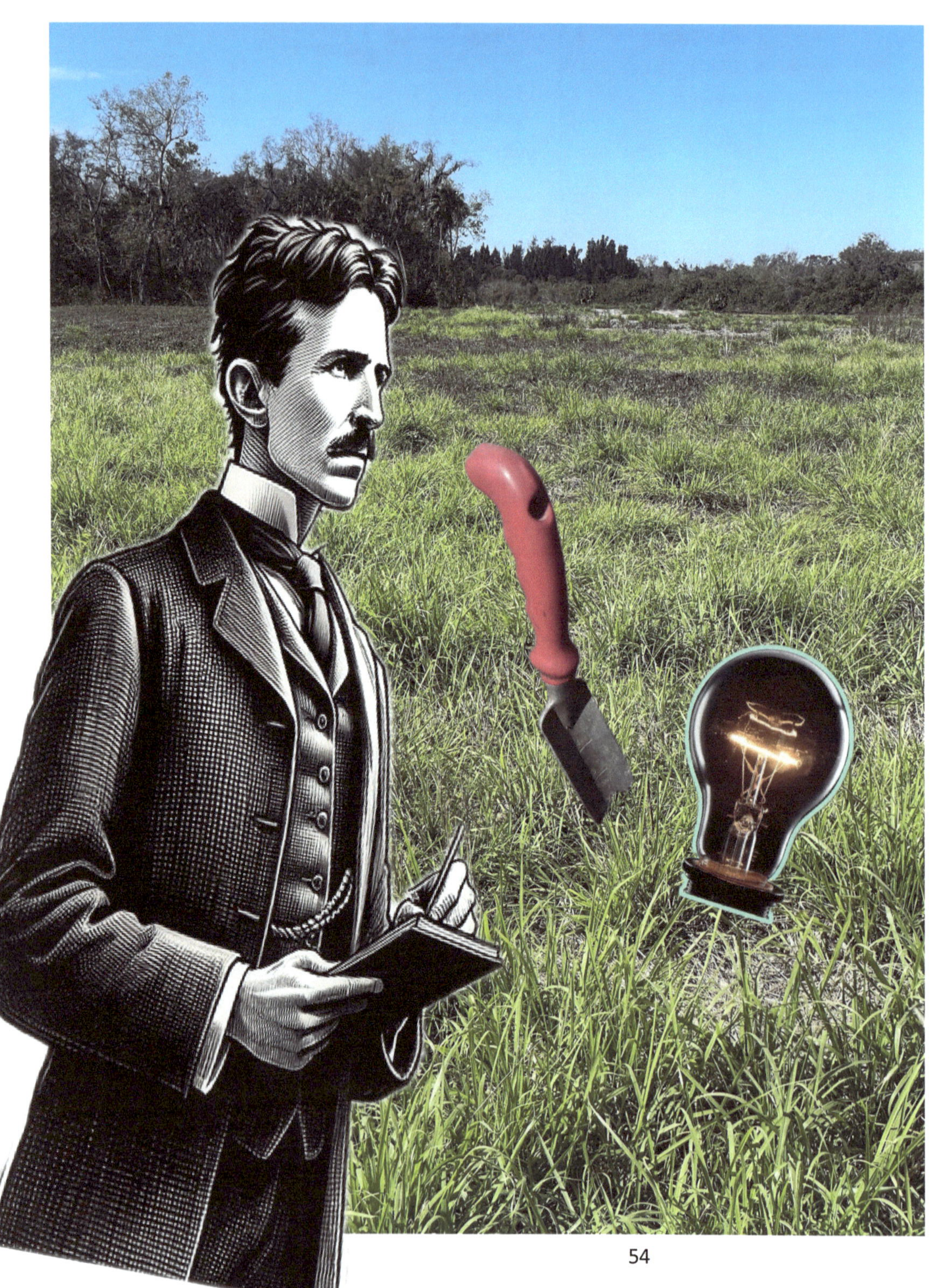

An early theory on adding new electricity generation that still might grow on you,,,

If you plant a light bulb in your garden, does it grow into a power plant???

Mother Earth News

A **power generation plant** in Florida with customer friendly solar panels in the parking lot just grew out of the landscape...

'Not In My Backyard'
Home values depreciate about 10% when view of wind turbines are less than 5 miles away.
The total loss in values across all US houses with a view of windmills adds up to a drop in the US of $24.5 billion

ELECTRIC VEHICLES

Trip is a new company headquartered in Columbus Ohio. Their focus is on Personal Mobility in the form of creation on an all new *e*-bike.

Rise of micromobility: The global micromobility market is worth about $180 billion today. McKinsey analysis shows that the value could more than double by 2030 to reach about $440 billion.

MicroMobility, driven by the Gen Z segment of our population is the major motivating factor, and of course it has to be powered by Eddie Current.

purposes, growing environmental concern for reducing carbon emissions, and increasing supportive schemes by the U.S. government encouraging the creation of bike highways which has been started in many cities are the key factors driving the market growth.

In the US, there are about 250 ebike brands currently selling bikes, in the EU, by comparison, there are nearly 500 ebike brands.

The total *e*-bike market in 2022 was in excess of $48 billion.

Rising demand for *e*-bike not only for recreational activities but also for daily commute

How an electric bike is classified:

Class 1: Top speed 20 mph; pedal assist only; allowed on bike paths and MTB trails in most states, with some restrictions

Class 2: Top speed 20 mph; pedal assist and throttle modes; allowed on bike paths and MTB trails in most states, with some restrictions

Class 3: Top speed 28 mph; pedal assist mode (to top speed), throttle mode (limited to 20 mph); allowed on streets and bike paths; off-road use restricted in most states to motorized trails only

Let me think,,,
At my age,
Should I try this
or not?

***Indecision* is the key to flexibility...**

4 Wheels are much safer, and at my age, safety is a major driver!

What kind of car does an electrician drive? A "Volts-wagon."

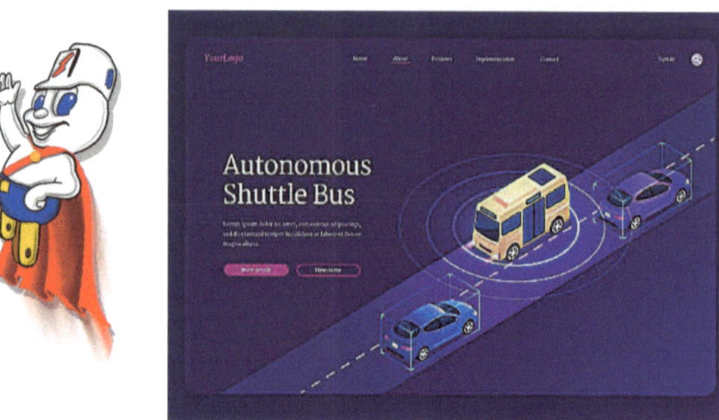

EV Manufacturing Facts:

BYD, a Chinese conglomerate, is the largest EV company based on the nearly 1.9 million EVs it manufactured in 2022, according to EV-Volumes. About half of those were plug-in hybrid EVs (PHEVs), and the other half were battery electric vehicles (BEVs).

Tesla built 1.3 million EVs in 2022, the most by a U.S. company and second most in the world. Tesla built the highest number of battery electric vehicles in 2022.

What is Electricity,,, & How Does It Make The Car-go???

Are the kilowatts that drive your new Tesla the same if they came from a coal fired power plant or from a source like a solar panel or wind turbine?
Lets ask Eddie Current:

"Well Mister Ed, here's my story. When a new car is delivered it has "a full tank," if you know what I mean. My buddies and I are stationed in all four tires just as you see us now. Then you give us some "juice," by stepping on the "go" pedal.

We are now ready to do our job. However, when you make us do a lot of work, we just plain run out of juice, we get too pooped to pop. I mean, how much energy would you have left over if you had to pull a 2 ton car another 300 miles on an empty belly? Even with a bunch of friends… We're no different then you…

Here is how we work. we hop *into* the tires and RUN round and round —*like a hamster on a treadmill,,, (engineers call this a motor)* thus the car goes forward, if in reverse, then I am recharging myself.

When we get tired, we need to be plugged in to a charging station which is merely giving us the energy we need to continue to do the work. Actually the 'battery' is really a lithium hot-tub, and plugging in to a charging station only heats up the hot tub which energizes us. But luckily for you, instead of needing beef, chicken, chicken thighs, chicken fat, salmon, sheep guts, and fish, we are happy to just suck the heat from the lithium.

Just a thought,,, When thinking back about the past in transportation imagine when everyone rode horses and only the rich had cars, and now today, everyone owns at least 1 auto and only the rich ride horses… imagine the future world where all cars were EVs, then along came the 'Internal Combustion Engines.'

Would they sell like hotcakes? Advertising would say: Here is a vehicle that is half the weight, half the price, that will cause less than 1/4 the damage to the roads, that could be refueled in 1/10 the time, with a range up to 4 times and had no limitations due to weather.

Maybe, most importantly, its manufacturing does not demand environmentally damaging non-renewable rare earth elements,,,

The battery pack (Lithium Hot-Tub) on the typical EV

Could this be,,, What are we hearing sounds like the shouting's of an aluminum siding salesman??

This year was shaping up to be a good one for the workers at the GM Orion Assembly plant in a working-class suburb of Detroit. After winning a pay raise following last year's United Auto Workers strike, they were slated to start production later this year on a marquee product for GM: the electric Chevy Silverado pickup truck.

But like thousands of other workers on the front lines of the electric vehicle transition, they have hit some bumps in the road. GM told the nearly 1,000 workers at the Orion plant in December that they were being laid off until late 2025 to make engineering improvements and amid cooling demand for electric vehicles. Their last paycheck was their holiday pay the week of Christmas, and many are still waiting to find out if they will be offered a job at another plant.

"It's been a very somber moment here the last month," said an employee at the plant who asked not to be identified because they weren't authorized by GM to speak to the media. "We thought we finally got a little bit of a break. I think the automotive industry is hurting. In my opinion, they put the carriage in front of the horse."

"Ultimately it all stems from demand, and demand is just not showing up to where all these CEOs thought. So a lot of the initial targets put out by GM or Ford a couple years ago have maybe proven to be a bit too optimistic and probably too aggressive," said Gabe Daoud, a sustainable energy senior analyst at TD Cowen. "I think everybody was expecting the entire car fleet to change overnight and go electric, but that's obviously just impossible and impractical."

Along with GM, Ford said in January it is cutting back on production of its electric pickup truck the F-150 Lightning amid slowing demand, eliminating 1,400 workers from the production line in Dearborn, Michigan, starting in April.

Someone in the Traffic Department has a great sense of Humor!

https://www.motortrend.com/reviews/2022-lucid-air-grand-touring-performance-yearlong-review-update-1-cold-weather-range-test/

What Is MPGe? The New Fuel Efficiency Rating for Hybrids and EVs

Here is an interesting fact:

Everyone knows what a gallon of gasoline looks like,,, how many times have you filled up a can at the gas station for the lawn mower or boat...but, what does a kilowatt hour of electricity look like??? How many "D" batteries does it take to make a kilowatt hour of electric power? This might be a featured topic for a 2nd Edition of "What is Electricity...and where does it go when it leaves the pump???"

Yes, You are correct, Eddie Current is running on a treadmill, just like a hamster. This produces the energy to power an EV.

It is also great training for running inside the tires. This is actually the activity which creates the torque that makes the EV go forward! When he turns around, then you have out the selector into reverse, and you are putting 'juice' back into the battery.

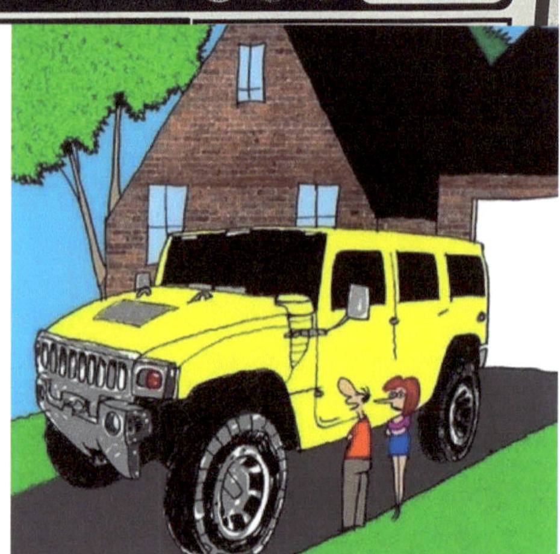

"It's actually electric. The only bad thing is you can only go 50 feet before you have to charge it again."

An EV takes 40% less labor to produce,,, think about that— only 60% of the labor to produce a more common gas powered vehicle. And In fact, according to Tesla, their drivetrain only has about 17 moving parts compared to the 200 or so in a typical drivetrain for an internal combustion driven auto.
Sooo, why the high cost of an EV??? Or better yet, Why the duplicity of our government in their subsidizing and pushing a 'Green Agenda' while, at the same time claiming to be supporting the labor organizations???

Everyone knows what a gallon is. We experience it almost daily. When we go to the grocery store and by a gallon of milk or a gallon of Arizona Tea. And, most commonly, when we fill up our old internal combustion engine cars the meter on the pump tells us how many gallons , down to the tenths, was pumped into the tank.

But now our government wants us to give them complete trust when they say that the energy in one gallon of gasoline is equal to 33.7 kilowatt hours of electricity. This is very close to the magicians slight of hand trick. Are we comparing to regular or premium gas? What octane rating? E85 or diesel?

Who has seen a kilowatt hour of electricity, let alone 33.7 of them in one place? Who do you trust? The savings and annual fuel cost are in the largest type font,,, while cost of a KWH is at the bottom where nobody ever reads that far down and is so small you can't even see it,,, $0.13 cents per kilowatt hour,,, *reality* only in your wildest dream.

With this seemingly sound criteria, the government, probably Democrats, created an evaluation terminology called MPG*e*, representing the number of miles the vehicle can go using a quantity of fuel with the same energy content as a gallon of gasoline," so it says on the EPA website.

Another chapter in the $8.8 trillion government Hoax!

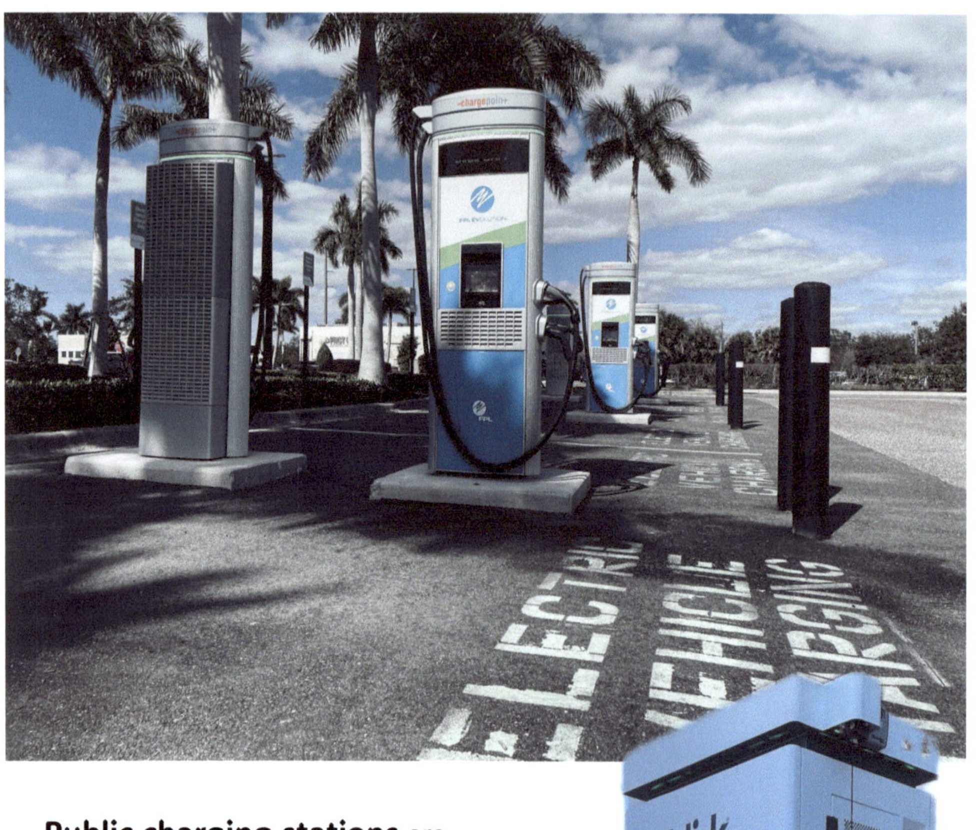

Public charging stations are becoming more readily available, many thnx to government sponsored incentives.

Tesla - Toasters - Teslas

Here's the latest news on **EV's in the US,,,**

Adding insult to injury for the EV names is news that is the
news that Hertz is selling 20,000 EVs, or one-third of its
total U.S. EV fleet, (because of high maintenance costs
and the inability to rent at reasonable prices) in favor of
purchasing gas-powered vehicles and incurring $245M of
depreciation expense in the process.

*...one-third of its total U.S. EV fleet, in favor of purchasing
gas-powered vehicles.*

January 11, 2024 12:39 PM EST) CFRA analyst Garrett Nelson

What do you call a Rooster
staring at a piece of lettuce???
A Chicken-Sees-A-Salad!!!

Crazy starts with a **"C",** and so does **C**alifornia... Several cities in California have now passed laws stating that the police cannot give traffic tickets to Electric Vehicles (EV's.) So, when you are in your Tesla, or Lucid, and you are in an accident and hear the sirens coming up behind you, your best bet is hop into the back seat! Your safe and the auto can't get a ticket!

We're happy you could contribute to our Policemen's Ball,,, Here's your coupon for a free bucket of Eddie Currents on your next fill-up,,,

I just got a full tank of gas for $22. Granted, It was for my lawn mower, but I'm trying to stay positive.

Electric Cars: A plan that doesn't work to fix a problem that doesn't exist.

Electric charger powered by diesel generator. We are the stupidest species on the planet...

⚙ · Rate this translation

Outdoor Temperature, -4⁰ F

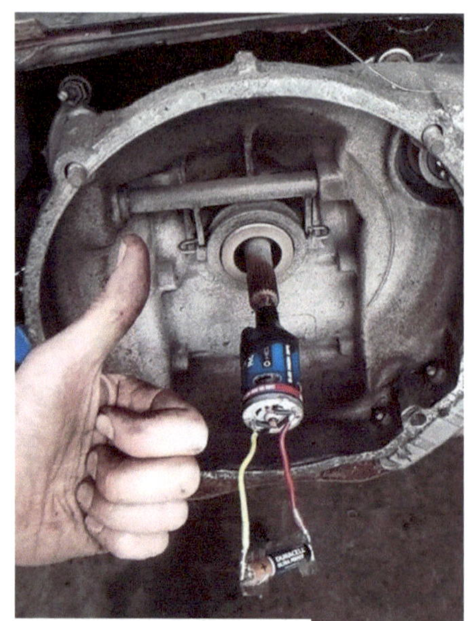

So when these Electric Car's batteries overheat and burst into flames, is it still considered a zero emissions car?
Asking for a friend.

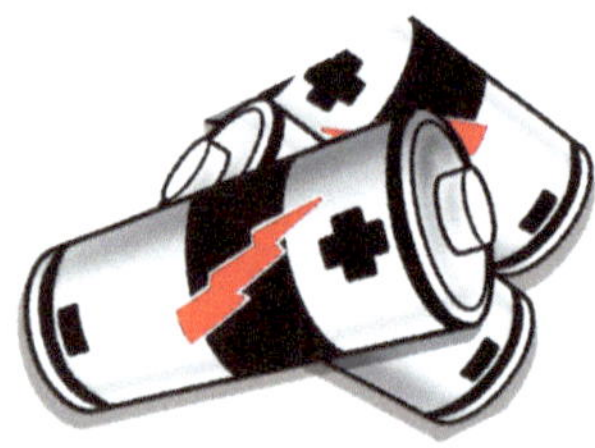

My day is complete.

A gas powered generator for an electric car.

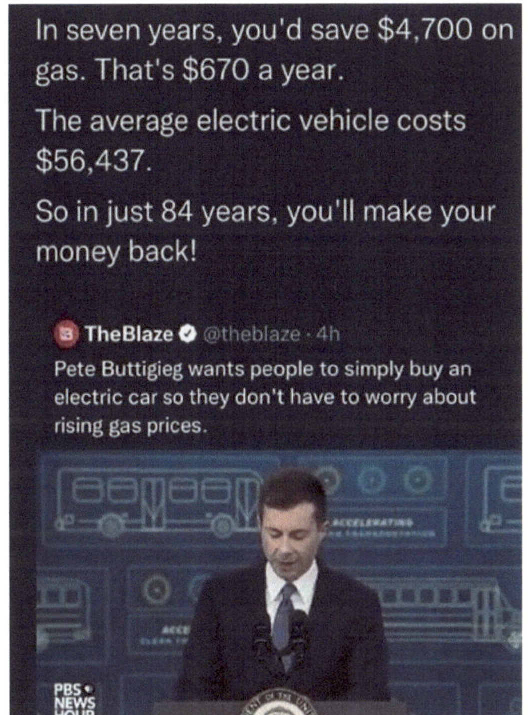

More than 284 million vehicles (cars, motorcycles, trucks, & buses) are registered in the US.
The number of cars in the world is estimated at more than 1.4 billion.
Europe has about 400 million.
And, reportedly, China gets the trophy as having the most autos (415 million).

How to write "I changed a lightbulb" on your resume:

"Single-handedly managed the successful upgrade and deployment of a new environmental illumination system with zero cost overruns and no safety incidents."

If you get sick riding in an EV, would that be diagnosed as an Auto-Immune disorder???

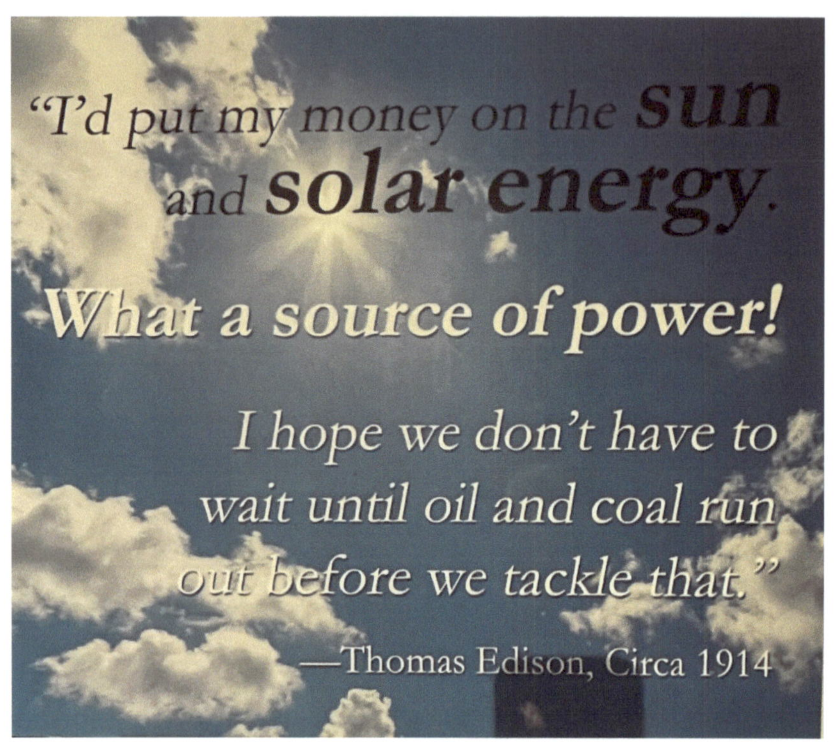

Here we are,,, into the mid-2020's,,, millenniums after these elements arrived in our universe,,, and they are still here,,, we have no shortage of the **sun, and tornados and lightning!**

And now, we have a whole bunch of people telling us how we should live our lives...

When I asked "Why," when I was a youngster, My grandfather often told me ~~~ *'If you don't understand something, then there is money behind it.'*

From an anonymous Wisconsin State Trooper

I'm not a fan of electric vehicles. Too many variables affecting battery consumption. Definitely not suited for cold climates. The following experience just cements my distaste for EV's, especially Teslas:

I get sent to a motorist assist the other day, at the start of our snowstorm. Tesla on the side of the interstate, dead battery. So, I arrive on scene and the occupants have the right-front door open. They tell me that they can't open any other doors, because the battery is dead. Sure enough. Can't open the doors from inside or outside. The driver also can't get her license out of the glove box where she put it during their trip. Because the glove box opens electronically... and the battery is dead. You actually have to use the computer in the center of the dash to open the glove box.

They said they had 10% battery left, should've been plenty to get from that location to the charging station nearby. Then all of a sudden, the whole car shut off and they coasted to the shoulder.

So now I have to find them a tow. No one wants to tow EV's. Finally found one company to do it. 8-mile trip to the charging station in Tomah. $1,000! Normal vehicle on the flatbed would've been $150.

So now we're at the Tesla superchargers. Guess what. Can't open the fine charging port because the battery is dead!!! The charging ports open, you guessed it, electronically!!! And we also can't open the doors now (had to close the one open door when it was loaded onto the wrecker). The owner's manual is in the on-board computer, but the battery is dead.

I got the occupants to a store where they'd be warm while calling the rental company to figure out how to charge this POS, so I'm not sure of the outcome. I had to leave for a crash report. EV's may be the way, someday, but certainly not today!! *I'll stick with my dinosaur burner.*

Pedal Power:

Yes, it is possible to generate electricity by pedaling a bike connected to a generator. This concept is known as human-powered electricity generation or pedal power. The principle behind this is converting mechanical energy generated by pedaling into electrical energy using a generator.

The electricity generated can then be used to power various devices or stored in batteries for future use.

Pedal power has been used in various applications such as charging small electronic devices, lighting up homes in remote areas, and even powering concerts and events.

It is a sustainable and eco-friendly way of generating electricity, as it does not rely on any external source of energy, such as fossil fuels.

However, it is important to note that the amount of electricity generated will depend on several factors such as the efficiency of the generator, the strength and endurance of the person pedaling, and the duration of pedaling.

Generating enough electricity to power large appliances or industrial machinery may require significant effort and time. Additionally, it is essential to ensure safety measures are in place when setting up a human-powered generator, particularly when dealing with high voltages.

The cost-effectiveness of electricity generated by non-fossil fuel sources depends on various factors such as the initial investment, maintenance costs, and the cost of fuel or energy required for generating power.

Generally, renewable energy sources such as solar, wind, geothermal, and hydroelectric are becoming increasingly cost-effective as technological advancements are reducing the initial investment and maintenance costs.

Needless to say, the cost of refueling this generation source might make this type of renewable on the very expensive side...

I bet Gold's Gym or Planet Fitness would attract customers if these were in their equipment line-up.

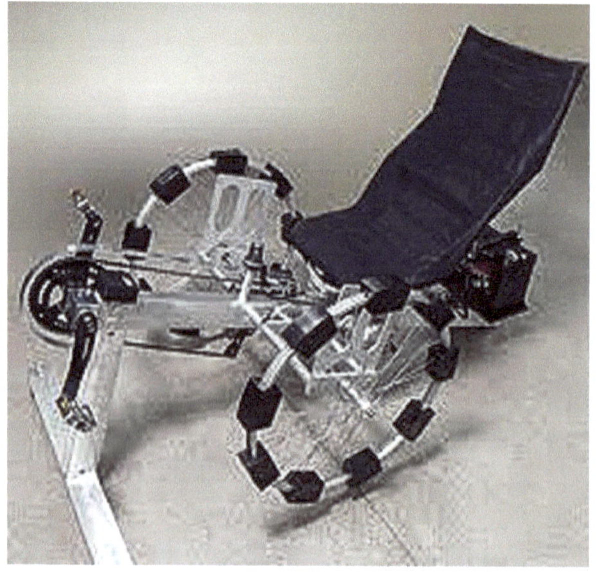

However, it is essential to note that the cost-effectiveness of electricity generated by non-fossil fuel sources can vary depending on the specific location, climate, and available natural resources.

For instance, while solar power may be more economical in areas with plenty of sunlight, wind power may be more profitable in areas with high wind speeds. Therefore, it is crucial to evaluate the specific costs and benefits associated with different renewable energy sources before making any investments.

Ready Kilowatt, a popular mascot for municipal electric companies in the mid-twentieth century, may not be able to become a superhero for electrification, but the concept of electricity itself has the potential to transform communities and even entire countries. In many parts of the world, access to reliable and affordable electricity is still a luxury, and millions of people rely on traditional, polluting sources of energy such as coal, oil, and wood for lighting, cooking, and heating.

By expanding access to clean and renewable sources of energy, we can improve health outcomes, create new economic opportunities, and reduce the harmful environmental impacts associated with fossil fuels. From small-scale solar installations in rural villages to large offshore wind farms powering entire cities, electrification offers a pathway towards a more sustainable and equitable future.

However, achieving universal electrification will require a concerted effort from governments, civil society, and the private sector. This effort will involve overcoming numerous challenges such as inadequate infrastructure, limited financing, and political barriers, as well as ensuring that energy systems are designed to meet the needs and priorities of local communities. Nevertheless, by working together, we can leverage the power of electricity to shape a better world for ourselves and future generations.

*Do you think
L A Fitness
should install
these in all
their Gyms!!!*

ELECTRIC PLANES

It's a Bird,

No,, it's a Plane,

Nooo,,, its an Electron!

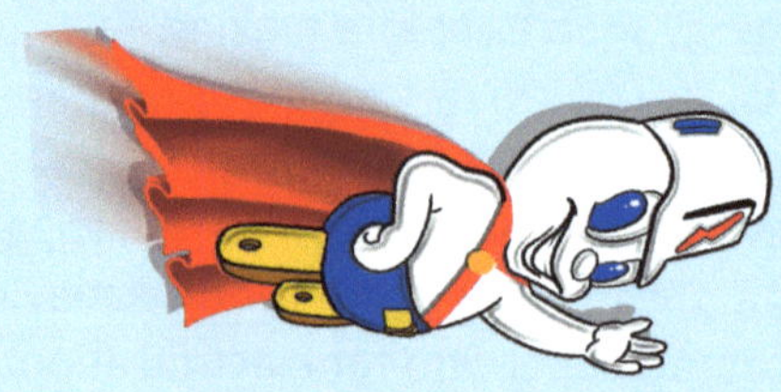

Private Jets:

Private Aviation today represents 2.5% of the world's carbon emissions (and likely much more in non-carbon emissions) yet only 1% of the world's population are responsible for about 50% of all aviation emissions. Moreover, private planes are up to 14 times more polluting, per individual, than commercial planes and 50 times more polluting than trains, according to reporting by Transport and the Environment.

So **going to an electric powered plane is, I guess, a natural progression** that we need to think about and develop. Afterall, planes do create a great deal of CO_2 especially those jumbo jets but, the heavy weight of those batteries has been a nagging problem. And the weight does not decrease as the flight goes on ~~ batteries' always weight the same at take-off as they do on landing~~~ (Eddie Current is weightless).

Today, more than 250 electric powered plane companies are currently developing worldwide. Some expert observers predict that that electric airplanes will be commonplace before the end of the next decade. You can always find the 'talking heads' that will verify everything you are trying to use to confirm beliefs.

Consider as a major significance the number of parts and component manufacturers. In the U S alone, more than 1400 of these support manufacturing businesses total the aviation industry.

Motivation includes: higher efficiency, quieter, safer, and, much GREENER! (much greener than aircraft relying only on traditional internal combustion engines). Lets understand that electricity already plays a very critical role in todays aviation. Electric motors drive many components necessary for control of the aircraft, and the avionics provide seriously critical functions in navigation, convenience and safety.

Electric motors were invented in the 1830's, and battery-powered cars were first invented and manufactured in the 1890's, however the next huge changes in aviation will not take hundreds of years but will happen in the next decade.

Following are just a few of those that have already developed and are currently filing for certification on various design strategies and models.

As we discovered in the previous chapter, mobility is one of the hottest sectors, with start-ups and traditional OEMs constantly developing new technologies and transportation options. *e-Planes* are certainly a sector that is gaining a lot of attention from investment companies and one of the disruptive trends and technologies that will shape the future of mobility and the impact that they will have worldwide.

Tesla Just Silently Filed Paperwork That Could Change Everything, I mean Elon not Nickola ~~~

Now, why didn't I think of this, a hybrid plane,,,

A hybrid electric plane is believed to have set a new record
for hybrid aircraft endurance during its flight over California
this past December. The Electric EEL from Ampaire, an
aircraft systems company, covered 1,375 miles and
flew for a total of 12 hours, with more
than 2 hours of fuel and battery reserves
remaining after its flight.

Drones can be used for knowledge acquisition,,,

Lets start with Drones, they are small, light weight, need only short periods of flight, don't need to carry heavy pay -loads and have most of the critical navigation technology already developed.

First, think about miniature Eddie Currents (electrons).
Are these electrons smaller, Lighter in weight, than their brothers that drive the huge diesel electric train engines?

Students can learn about various drone operation technologies, including aerodynamics, aviation, programming, and data analysis, by working with drones. Drones serve as a tangible tool that encourages exploration and problem-solving.

Drones are basically Flying or floating Computers powered by Eddie Current.

Military Drone Functions & Modern Warfare Impact,,,

These unmanned aerial vehicles have revolutionized warfare through their applications in intelligence gathering, target acquisition, precision strikes, force protection, surveillance and reconnaissance, search and rescue, logistics and supply, and combat operations.

Today, they are fossil-fuel powered, however, who know what programs are in the works?

Lilium (a California based firm) is the first vertical take-off and landing jet. Its proprietary technology is termed **Ducted Electric Vectored Thrust.**

Jet engines are integrated into the wing flaps and provide advantages in aerodynamic efficiency, payload, and a lower noise profile.

It is now entering a phase of component testing. The first conforming aircraft is expected in May and battery testing is also scheduled, ahead of conforming aircraft production in Q3/4 2023. Lilium declines to discuss delivery timelines but confirms EASA certification flight trials will begin soon after and certification is expected in 2025.

The Lilium uses motors that develop 100kW. The maintenance data we have from automotive experience suggests the only time you'd need to remove the motor is after a bird strike or similar incident. It will fly forever.

What's a Whack???
How can I be out of it???

Vertical lift and cruise power comes from multiple electric motors driving ducted fans. Each unit is effectively a jet engine. A turbine, or small-diameter fan, forces air through a narrowing duct, where it is compressed and exits via a variable nozzle. Mounted in clusters on hinged wing and canard surfaces, the jets move as one in the transition from vertical to horizontal flight.

Aircraft design is all about compromise. The helicopter's vertical takeoff and landing (VTOL) capability, for example, comes at the cost of a draggy speed and range limiting main rotor. There is also the irrefutable fact that taking off vertically requires huge amounts of energy.

Considering contemporary battery energy density, an electrically-powered VTOL aircraft is further compromised, since high energy consumption in vertical flight strictly limits its range. Lilium's eVTOL concept, called Lilium Jet, aims to optimize the compromise.

Ducted fans or propeller?

VTOL, Vertical Take-Off and Landing, has emerged as a major criteria for this new category of electric powered planes, but why do they need 6 motors and propellers when a standard helicopter has only 1 huge propeller.

JOBY is another venture headquartered on the west coast that is developing an electric powered aircraft to function as an air taxi in major cities. Powered by six electric motors, the aircraft takes off and lands vertically, giving the flexibility to serve almost any community.

Today, a team of more than 1500 passionate engineers, experts, and leaders, are all focused on bringing this pioneering vision to life.

Headquartered in a world-class manufacturing facility in Marina, CA, Joby also has offices and workshops in Santa Cruz, San Carlos, Washington, D.C., and Munich, Germany.

AutoFlight's Prosperity eVTOL air taxi flew autonomously between the Chinese cities of Shenzhen and Zhuhai.

AutoFlight is designed as a five-seat air taxi—created for a pilot plus up to four passengers. Its recent flight, in the Hong Kong area, was fully autonomous.

This Chinese version utilizes separate propellers for lift and forward propulsion.

ELECTRICITY STORAGE

'EverReady Eddie' is a close relative to Eddie Current

Much like refrigerators enabled food to be stored for days or weeks so it didn't have to be consumed immediately or thrown away, energy storage lets individuals and communities access electricity when they need it most—like during outages, or when the sun isn't shining or the wind is not strong enough to turn the giant blades. Storage can reduce demand for electricity from inefficient, polluting plants that are often located in low-income and marginalized communities. Storage can also help smooth out demand, avoiding price spikes for electricity customers.

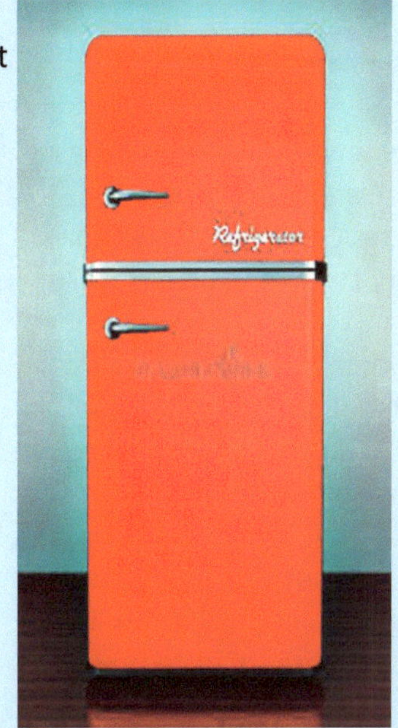

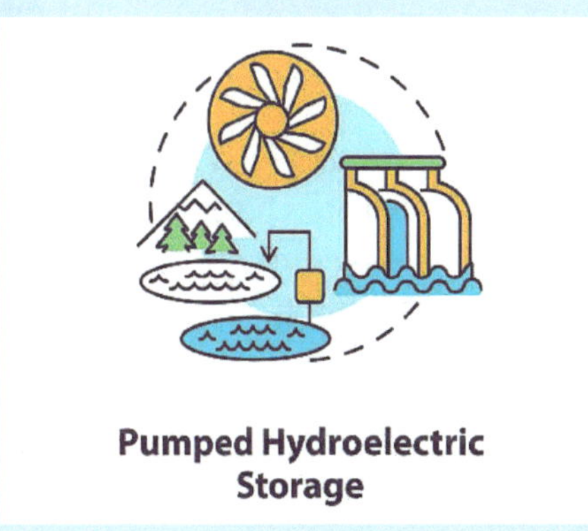

Pumped Hydroelectric Storage

Battery storage of electricity

Where does this chapter begin???

Ah, non battery storage of electricity. Now, there's the age-old debate about electricity and its practical uses. While batteries can certainly be a great tool for storing electricity, many people are unaware of the other, less conventional methods of storing energy. So, why not have some fun with this topic and explore the humorous possibilities of non battery storage of electricity?!

Let's start with the most obvious option: Charging up chickens - yes, it is exactly what it sounds like! We all know that chickens love to scavenge around and peck at things, so why not harness their energy by making them charge electric cars with their beaks? It's a simple concept - just attach the car's output port to the chicken's neck and make sure they keep pecking away. You'll soon see your car's energy level increase in no time!

If chickens aren't quite your thing, then what about using the force of a river or a stream to generate electricity? Picture it: attaching powerful turbines to a body of flowing water and watching as electric current is produced from it. What makes this even funnier is that adding more water flow will only increase the rate of electricity generated, which means you can shout 'Give me more power!' to the river - and it just might respond.

Finally, let's not forget the wonders of wind power - who said you need to attach turbines to a particular building to reap the benefits? Instead, why not use individual kites and hang them off of tall buildings? When the wind blows, it will cause the kite to lift, thus engaging the turbine to rotate and generate electricity. It's an ingenious way of harvesting the energy of the wind while also having some fun with colorful kites in the sky.

Ultimately, non battery storage of electricity can bring many unexpected surprises - and laughs! Unfortunately, lithium-ion batteries are not the only way to store electricity. While they remain the most popular choice, there are a number of alternatives that offer different advantages.

For instance, one approach is to use chemical reactions. These reactions create saltwater solutions, which can then be stored in a tank or other container. Over time, the liquid will naturally break down into its basic components, releasing energy in the process.

Another option is to use flywheels; these are mechanisms designed to store energy in the form of motion. Generally speaking, the flywheel is attached to an electric motor, which then stores rotational energy. This works great for applications that require frequent bursts of power, as the flywheel can quickly provide it.

In addition, some researchers are experimenting with using the power of smell to store electricity. Specifically, specialized sensors can detect the smell given off by flowers and convert it into electrical energy.

Let's start this chapter with lithium...

When the world's most valuable lithium

company last year announced plans for a $1.3 billion plant in South Carolina, local officials hailed it as transformative for the Palmetto State.

As someone once said,,, Is this the rest of the story??? Please check out the page with the story titled:

Disaster Capitalism

The high-tech project from Charlotte, N.C.-based Albemarle was designed to process different sources of lithium, including from recycled batteries, and serve as a supplier of the critical mineral for South Carolina's burgeoning electric-vehicle industry.

Less than a year later, those plans have been hobbled by a crash in battery metal prices, undercut by a slowdown in electric-vehicle sales growth in the U.S. and China. Albemarle has deferred spending on the project, amid companywide cost-cutting that includes layoffs and delays to other investments as well.

Producers of lithium and nickel, which are used in lithium-ion batteries for EVs, have been stalling projects and closing mines to save cash after a painfully quick fall in commodity prices. Prices of lithium are down as much as 90% since the start of last year, while the price of nickel has roughly halved.

*Please refer to **Disaster Capitalism** on page 35*

*So much for this background info.
Now, let's start at the
beginning on the next page,,,*

ELECTRIC MERCURY INDUSTRIAL TRACTOR

ca. 1917, Mercury Manufacturing Company, Chicago, Illinois

Powered by thirty Edison storage batteries, this tractor was used for over 55 years at the Elgin Corrugated Box Company in Elgin, Illinois. Each battery was made of steel and iron, with nickel hydrate and iron oxide in an alkaline electrolyte. Edison's storage battery was manufactured from the early 1900's to the early 1960's.

Above: A Mercury Industrial Tractor in use (date unknown).

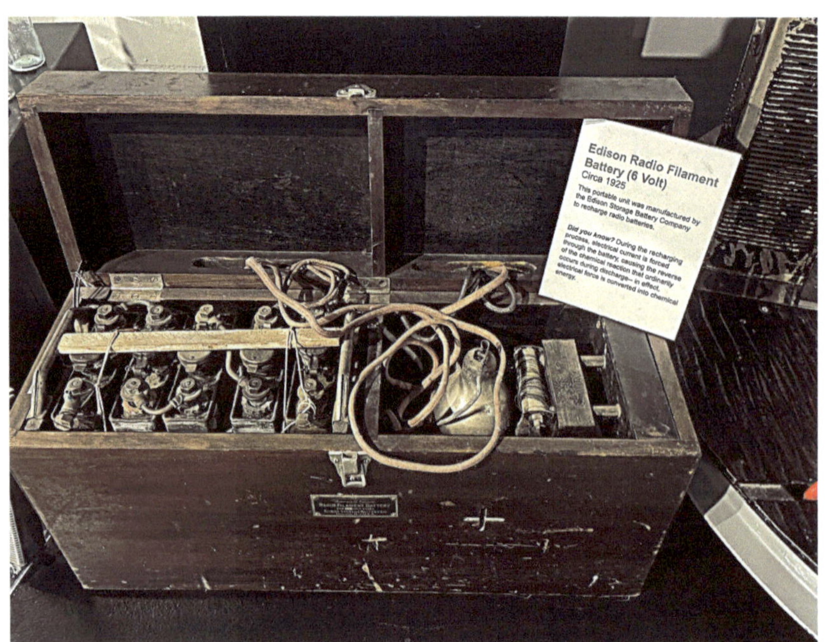

60 years on nickel hydrate—iron oxide batteries! My golf cart batteries don't last that long…

Did all this start with Thomas Edison???

From the winter estate of Thomas Edison in Fort Myers, Fl…

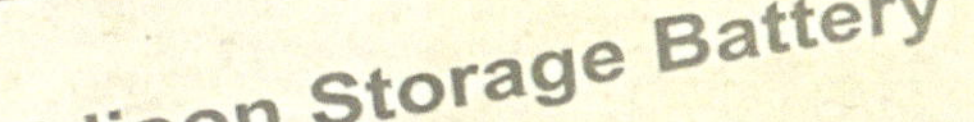

Edison Storage Battery
Circa 1930

This nickel-iron battery has thick glass to hold the potassium hydroxide electrolyte. Batteries such as these were highly valued by the railroad industry due to their reliability and capacity to hold a charge for a long period of time. This particular battery was used to operate rail switches.

"*Whether you think you can or think you can't, you're right.*"

Henry Ford

161 U. S. Patents

A view of the Edison lab and work-shop in Florida.

Edison Nickel-Iron Alkaline Battery
Circa 1910

In 1899 Edison began conducting extensive research on secondary batteries in hopes of finding a viable alternative to the existing lead-acid batteries which were heavy, difficult to recharge and corroded quickly. His idea was to produce a new and improved storage battery, one that would be strong and stable enough to propel an electric vehicle. Edison's research with alkaline cells led to the eventual development of a Nickel-Iron-Alkaline storage battery that he claimed would "run for 100 miles or more without recharging." His new batteries were designed to be used for industrial purposes such as electric delivery trucks, submarines, mining lamps, as well as train lighting and signaling. The Edison Storage Battery Company became one of the most successful divisions within Thomas A. Edison Inc. during the 1920s and 1930s.

Besides generating electricity, a **giant balloon filled with Liquid air** can have many other applications. For example, this technology could be used for cooling or heating purposes. The liquid air could be stored in the balloon and then released or pumped into buildings or homes to help maintain the optimal temperatures. Additionally, it could be used to generate renewable energy in certain areas that are not connected to a grid. Furthermore, it is possible to use this technology to store energy produced from solar panels, wind turbines and other sources of renewable energy. It also has potential to be used as an energy buffer, to store excess energy from these sources and then release it when needed. Last but not least, it can help to reduce emissions, since energy can be generated without burning fuel!

Yes, it is definitely possible to use a giant balloon filled with liquid air to power a city! The balloon stores energy that can be released in the form of electricity when needed. This energy can be used to power homes, businesses, and more. Additionally, with advances in technology, the process is becoming even more efficient and cost-effective. So while it may have seemed like a far-fetched idea not too long ago, it is now becoming more and more realistic.

Not these ballons...

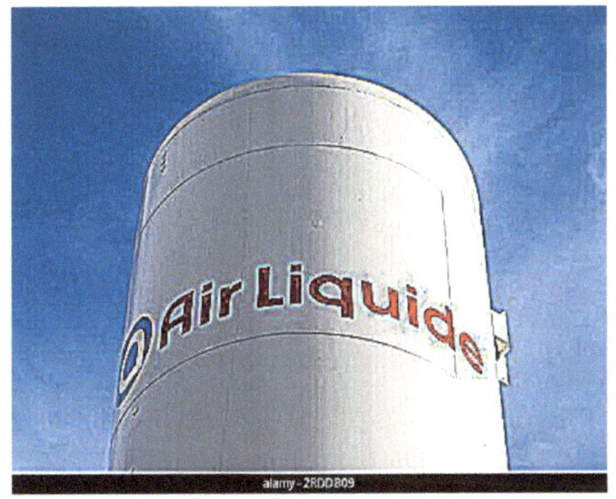

Liquid air batteries and gravity storage are among a number of solutions.

Liquid air energy storage is a fascinating topic, and one that has a lot of potential to be used in energy generation!

One funny story about this technology is that it involves using a giant balloon to store energy! In essence, the balloon is filled with liquid air and is pressurized, giving it the ability to store energy. The energy is then released when the balloon is "tapped" to have some of the liquid air evaporate, producing electricity in the process! Pretty cool, huh? Maybe not!!!

We have all experienced how wind and solar power have gone from being exotic technologies to common pieces of engineering that are seemingly competitive with fossil fuels. Fossil fuels retain what edge they have only because of their reliability.

To overcome the reliability requires cheap grid-scale energy storage that can be scaled up to match generation scale. Today, the market leader is the lithium-ion battery, the workhorse of applications from mobile phones to electric cars. They are reliable, scalable and well understood.

Most proposed alternatives are clumsy, poorly understood, unscalable or all three. But there is one that, because it relies on putting together pieces of engineering used routinely elsewhere, and thus proven to work, might give lithium-ion batteries a run for their money: liquid air.

But, here is a real solution: At a temperature of -196°C, all of air's component gases will liquefy. Doing this is a routine, electrically driven industrial procedure. Storing liquefied gases in bulk is also a routine piece of engineering. The result occupies a 700th of the volume of those gases at room temperature. Then, when liquid air is warmed and allowed to expand, it does so forcefully.

Using a device called a Dearman engine (after its inventor, a Briton named Peter Dearman), that forceful expansion can be employed to spin turbines, and thus generators, thereby recovering part of the electricity used to liquefy the air in the first place.

At the moment, the plant in England recovers as electricity just over half of the power used to liquefy the air in the first place. With engineering tweaks, efficiency will be increased.

One benefit is to capture, probably in oil or salt, the heat generated as the air is compressed prior to its liquefaction. This will aid in increasing the power output to the same figures achieved by the lithium batteries.

Eddie Current is a graduated Undercover
Classified Training Instructor

Hybrid Power Plant:

Here is a comprehensive package that integrates (engine power plant and energy storage) seamlessly with renewable generation, ensuring stable and uninterrupted power generation, resulting in both economic and environmental benefits.

By attaching a turbine to a kite, you can generate electricity simply by letting the kite soar in the sky.

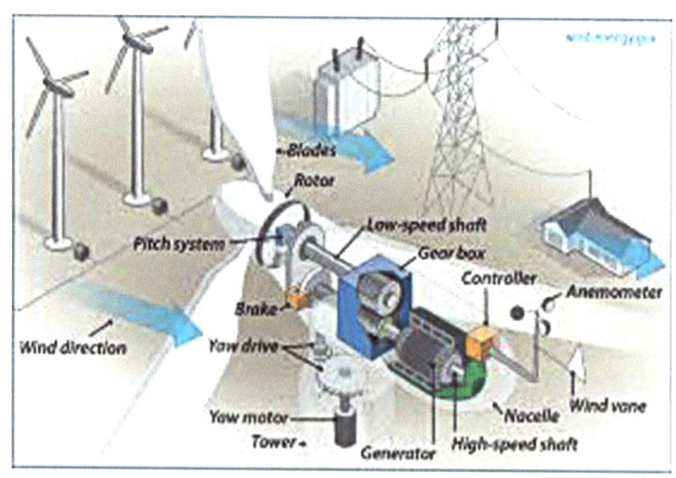

Compressed air energy storage (CAES) is a form of energy storage that uses compressed air to store and retrieve energy. It's a promising technology for storing excess energy that could otherwise be wasted. The funny side of CAES is that it can be used to power roller coasters and other amusement park rides, which adds to the entertainment factor of the energy storage system. Additionally, when using CAES, the same air is recycled, making it more sustainable and cost-effective than other forms of energy storage. Finally, CAES can provide ancillary services such as frequency regulation and black start capabilities, making it an attractive option for utilities

Compressed air energy storage (CAES) is a form of energy storage that uses compressed air to store and retrieve energy. By compressing the air in tanks or cavities and storing it until it is needed, CAES can provide ancillary services such as frequency regulation and black start capabilities, making it an attractive option for utilities.

In addition, CAES can power roller coasters and other amusement park rides while still being sustainable. This is because when the stored air is released and converted back into electrical energy, the same air is recycled and reused, reducing the amount of energy wasted. Furthermore, the process of compressing the air is often powered by renewable sources such as wind or solar, making the CAES process overall more sustainable than other forms of energy storage. Finally, its portability makes it an even more efficient and reliable source of energy.

Overall, compressed air energy storage is an efficient and sustainable form of energy storage that can provide numerous benefits, including powering roller coasters and other amusement park rides.

Compressed air energy storage (CAES) could theoretically be used to power a rocket to Mars, however it is likely not the most suitable option. The main reason being that rockets require a significant amount of energy to reach Mars, which CAES is unlikely to be able to provide since it is an "intermittent" form of energy storage.

It would be more practical to use a more consistent and reliable form of energy storage such as cryogenics or chemical fuels. These types of energy storage are not intermittent like CAES and can provide a more consistent output of energy for longer time periods, making them much more suitable for powering a rocket to Mars. It is also worth mentioning that the weight of the energy storage unit will have an impact on the overall efficiency of the rocket. Therefore, lighter forms of energy storage such as chemical fuels are usually preferred for applications involving rocket propulsion.

SGES Large-scale energy storage technology is crucial to maintaining a high-proportion renewable energy power system stability and addressing the energy crisis and environmental problems. **S**olid **g**ravity **E**nergy **S**torage technology is a promising mechanical energy storagetechnology suitable for large-scale applications. The basic concept of SGES a bibliometric study between 2010 and 2021 first introduced to show SGES technology's evolution and predict future trends. Various SGES technologies have been intensively investigated in equipment, principles, materials, progress, and mathematical models.

Compared with other large-scale energy storage technologies, SGES has many advantages: high cycle efficiency (80 %–90 %), large energy storage capacity, good geographical adaptability, and economy.

An essay on gravity-based energy storage might focus on current trends and innovations related to using gravitational forces to generate electricity, include weight, hydroelectric and tidal power. It could explore the pros and cons of gravity-based energy storage in comparison with other renewable sources, and discuss the potential applications in the future. The essay could also touch on the traditional forms of energy storage associated with gravity, such as dams, water tanks, and stone quarries.

Gravity Power returns energy to the grid at about 4¢ per KWh, less than half the cost of lithium ion, including the cost of energy lost in the round trip

Gravity battery: a type of electricity storage device that stores gravitational energy

In a gravity battery, the energy stored in an object resulting from a change in height. Height can also be called potential energy. A gravity battery works by using excess energy (usually from sustainable sources) to raise a mass to generate gravitational potential energy, which is then lowered to convert potential energy into electricity through an electric generator. One form of a gravity battery is one that lowers a mass, such as a block of concrete, to generate electricity. The most common gravity battery is used in pumped-storage hydroelectricity, where water is pumped to higher elevations to store energy and released through water turbines to generate electricity.

that used gravity to power mechanical movement was the pendulum clock, invented in 1656. The clock was powered by the force of gravity using an escapement mechanism, that made a pendulum move back and forth. Since then, gravity batteries have advanced into systems that can utilize the force due to gravity, and turn it into electricity for large scale energy storage.

Also known as Pumped-Storage Hydroelectricity (PSH) system, first pioneered in 1907 in Switzerland.

The Connecticut Electric and Power Company, in 1930, brought this concept to the US. Since then, gravity based pumped-storage electricity grew to the largest form of grid energy storage in the world.

The earliest form of a device

Pumped Hydroelectric Storage

PUMPED HYDROELECTRIC STORAGE

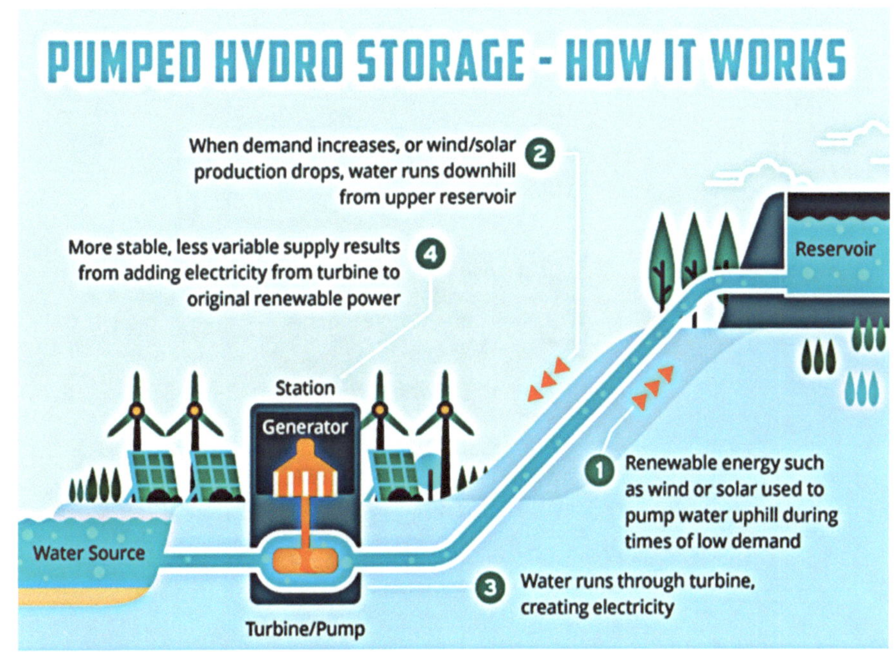

https://www.ucsusa.org/resources/how-energy-storage-works#:~:text=Flywheel%20Energy%20Storage%20Systems%20convert,(up%20to%2060%2C000%20rpm)

ENERGY ALTERNATES

That last chapter might have bored a few of you guys, Now, here is a page you can share with your friends and impress your kids.

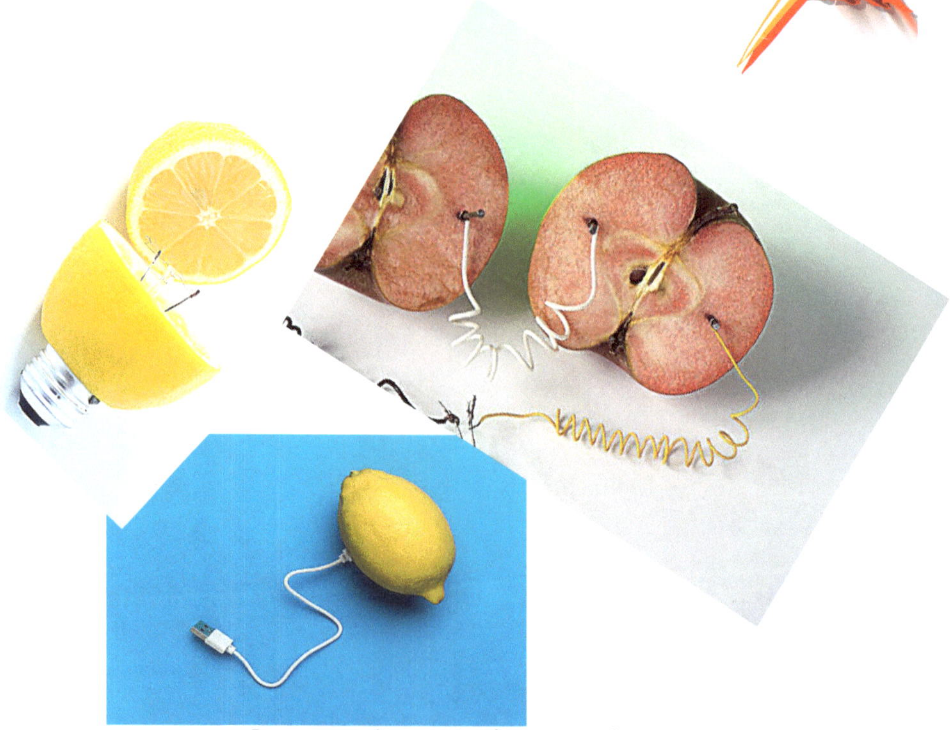

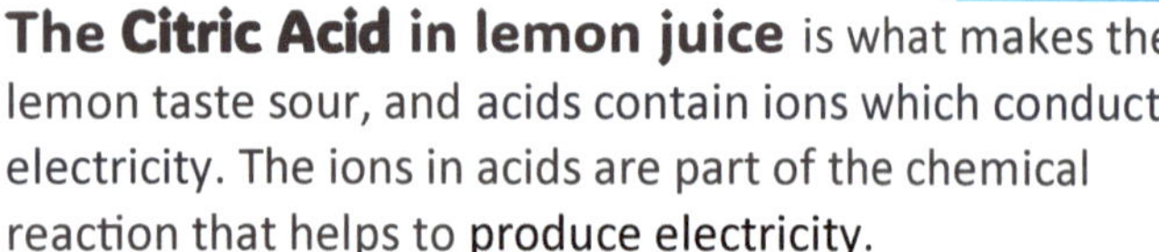

The Citric Acid in lemon juice is what makes the lemon taste sour, and acids contain ions which conduct electricity. The ions in acids are part of the chemical reaction that helps to produce electricity.

When you insert a piece of zinc metal (such as a galvanized nail) and a piece of copper (such as a penny) are inserted into a lemon and connected by wires.

Power generated by reaction of the metals is used to power a small device such as a light-emitting diode (LED).

By connecting a wire between the two electrodes, electrons are free to flow from the zinc towards the copper electrodes and create a current as well as voltage difference.

Surprisingly a potato can produce approximately one volt while a lemon can produce approximately 0.7 volts.

Another use of the potato is to use the juice and turn it into vodka… *Bet you didn't think you could get culinary lessons here also.*

~~~Watch out for the wind turbine blades! ~~~
DO NOT DISTURB
I NEED A SIGN THAT SAYS,
"ALREADY DISTURBED, PROCEED WITH CAUTION!
~~~

One of the most hilarious ways to generate electricity using the force of gravity would be to build a massive swing set. This swing set would have ropes attached to a counter weight on either side and as people swing, the motion would cause the counter weights to rotate. That motion in turn could be used to generate electricity. It would be both amusing to watch, and also provide a useful renewable resource.

Additionally, you could construct wind turbines with fun shapes, such as animals or other popular cartoon characters. As the wind blows, they would spin, offering a visually pleasing way to generate electricity. Finally, you could install an upside-down water slide that collects kinetic energy when someone slides down it.

This could be used to generate electricity, while providing a fun and humorous experience at the same time.

Many creative ideas have been proposed to make the process of generating electricity more fun.

One idea is to incorporate art into wind turbines. This could involve painting the turbines or adding sculptures and lamps to them, transforming them from utilitarian objects to beautiful works of art.

Another way to make the process of generating electricity more fun would be to create virtual reality (VR) simulations that teach users about renewable energy. These simulations could allow people to build their own wind turbines, solar panels, and hydroelectric dams, and explore the science behind how these systems generate electricity. Additionally, games and activities can be created to help educate people on the various forms of renewable energy generation. Finally, interactive art installations can be constructed using renewable energy technology, such as solar-powered sculptures that light up when it gets dark. Such installations would both be visually stimulating and informative at the same time.

Here's a project you can start in your backyard.

Scientists discover bees generate electricity!

Bees, those lovable buzzing creatures that fly around, are usually not credited with being able to make electricity. But, believe it or not, these industrious and busy little bugs are actually capable of doing just that!

This strange phenomenon has been studied by scientists who have discovered that bees can generate electricity through the process of *frictional induction*. When a bee moves along a surface, not only does it generate a small amount of static charge, but it also creates tiny electrical currents.

When a bee lands on a flower, the flower's petal provides a path for the current generated by the bee while it is pressing against the petal. As the bee moves along the petal, and more of its energy is transferred to the flower, the flower generates an electrical current. This current is what allows them to get their food--nectar and pollen--from the flowers.

The stem then conducts the electrons into the ground where they are stored as we found out earlier.

So, how can all this electricity help us humans? Well, the most obvious use of the electricity created by the bees is the production of honey. Today we can see that the electricity created by these little creatures could bee strong enough to power not only their hives, but also there could bee enough left over to fire up a TV and recharge a smart phone.

This type of energy production is renewable, meaning it can be used without harm to the environment or depletion of resources.

So, the next time you see a bee buzzing around your yard, remember that it might be creating enough electricity to power your home some day! Who knows, maybe the future of electricity will depend on the tiny buzzing of bees!

This concept beats the idea of training bees to tap dance on solar panels for generating energy but it would require a tremendous amount of resources and effort to train bees for this task, as well as ensuring their safety and well-being.

If bees were in charge of the energy grid, there would be several key differences compared to our current energy infrastructure.

First, it would be much more sustainable. Bees generate electricity through photosynthesis, converting the energy from the sun into an electrical current. This type of energy production is renewable, meaning it can be used without harm to the environment or depletion of resources. It would also be cheaper than electricity generated from traditional sources such as coal and natural gas, since it requires much less energy to produce.

Second, it could lead to improved air quality. By producing electricity through a renewable source, we can reduce the amount of pollutants released into the atmosphere from burning fossil fuels. This will lead to improved air quality, resulting in less respiratory illnesses.

Yes, other renewable sources of energy, such as solar and wind power, are usually a much better option than the idea of training bees to tap dance on solar panels for generating energy. Renewable sources of energy such as solar and wind power usually offer much higher rates of energy production and require less cost and effort than training bees to tap dance on solar panels. In addition, renewable sources of energy are also much more reliable, with consistent and long-term energy supply compared to training bees which can be unpredictable.

For example, solar power can be tapped during the day time when the sun is out, and wind power can be tapped during the night time when there is wind. Thus, these renewable sources of energy can provide a more consistent and reliable source of energy as compared to the idea of training bees.

No, unfortunately, it is not possible to get bees to generate enough energy to power a small city. Although bees are efficient and effective pollinators and have been used in some applications for generating energy, the level of energy production is still far too low to be able to power a city.

In addition, it would require a tremendous amount of resources and effort to train bees for this task, as well as ensuring their safety and well-being. Moreover, there are certain environmental risks with the idea of training bees to generate energy, as the fields and flowers used to feed the bees usually require pollination as well.

Today, renewable sources of energy such as solar and wind are much more feasible, effective and affordable options for powering a city compared to the idea of training bees, but, technology has a way of catching up once an intrapreneur gets involved!

The claim: Climate change is a 'scam' because Australia's trees and mangroves absorb more CO$_2$ than the country produces,,,

A July 15 Instagram post purports that one person's explanation of Australia's net carbon emissions proves that climate change is not real.

"Man gives crazy theory that proves climate change is a scam," reads the post, which includes a video of a person talking.

"According to government figures, Australia is producing 499 million metric tons of carbon dioxide per year," the speaker says, according to a text overlay on the muted video. "But what I also found is that one mature tree will absorb 48 pounds, or 21.77 kilograms, of carbon dioxide per year. And Australia currently has 24 billion standard size trees, absorbing 453 million metric tons of carbon dioxide per year. And that's just trees – doesn't include any other plant life – and particularly doesn't include mangroves, and ***mangroves will absorb 50 times more carbon dioxide per year than a standard sized tree***."

(Imagine Florida with more than 1300 miles of coast and all those mangroves.)

USA TODAY July 27, 2023

The **Daintree in Australia**, is a national icon. It's home to the oldest tropical rainforest on earth, estimated to be about 130 million years old. It's also Australia's largest rainforest, incorporating an area roughly 1200 sq km, and supports a staggering selection of animal and bird life.

This historical footnote is significant.

First, in the past millions of years, Daintree has contributed trees, plants, and most probably dinosaurs and Mother Nature, together with Father Time, has converted into fossil fuels that we are using today.

Second, the trees today are helping maintain energy independence by absorbing tons of CO_2 each year.

Third, this rich natural eco-friendly home to a large colony of bees, which we have already seen as a potential new major renewable energy source.

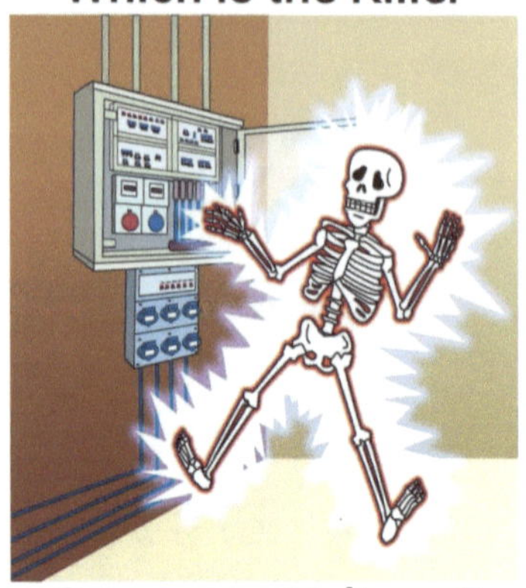

Would this be an oxymoron???

"Let's design of an electric chair with ultimate comfort as the main criteria…"

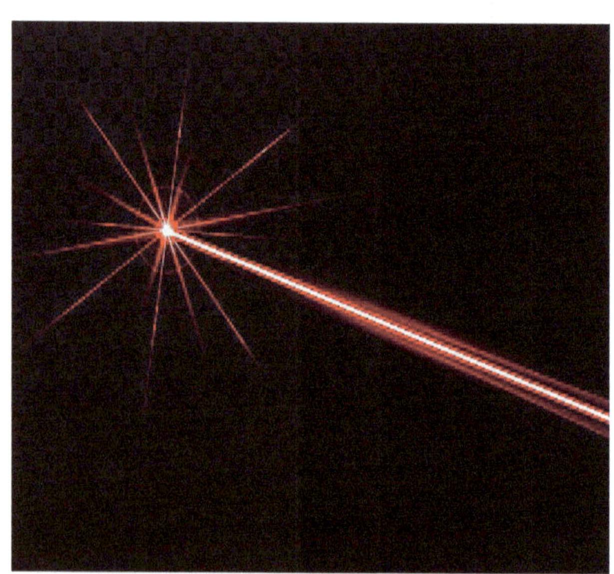

Also, Consider the Laser:

A laser beam can be a delicate, precise instrument used in surgery. Its flight path is very straight and will not droop or sag.

This could teach us that electricity can be a very powerful force, and we must never use it to hurt others, unless we need to teach them an important lesson.

How to turn your car alternator into a wind generator...

If you want your own wind powered generator, just check out Mother Earth News. There you'll find all you need is a vehicle alternator with a built-in voltage regulator, a fan and clutch assembly, some pipes, a tower or pole, some wire and a battery that you can get off your golf cart.

Parts is Parts... and will always have a place for the creative mind.

Who said the transition from fossil fuels to renewables would cost trillions and take decades???

Ponder this for a moment. 🤔
The wind farm in Mt. Pulaski has been running for 3 1/2 years. They have been replacing the generators in all the wind towers. There are 100 of them in this wind farm. So evidently the life span on the generators on these things is about 3 to 4 years. It takes 12 semi trucks and trailers, A 9 axle 500,000 pound crane, A 100,000 pound crane and 12 pick up trucks to change each generator. That is a huge amount of diesel fuel being used to maintain these wind towers. And the "Green Groups" would like You to believe they are all fueled by magic fairy dust.

What we never heart about is the **maintenance budget** of 'Going Green'… and with all these wind and solar farms, we are starting to realize the actual costs of the new green deal…

The US Transportation Department is investing $148.8 million for repairing or updating nearly 4,500 electric vehicle charging ports in 20 states.

Pocket change compared to the $7 trillion already spent.

More than electrical

fantasy! Its really magic fairy
dust,,,

Electricity Generation

Electricity generation can be a very boring topic. But don't worry, I'm here to inject some humor into it!

Let's start by talking about renewable energy sources. Solar power is one of the cleanest energy sources around. It's like a giant sunbaked pizza on your roof! Wind turbines are also great for generating electricity. They look like giant pinwheels in the sky, and make a great addition to any landscape.

How about non-renewable energy sources? Nuclear power is one way to generate electricity, but it requires a great deal of safety as well as proper waste disposal. In other words, you have to keep an eye on your radioactive rods at all times! Likewise, burning fossil fuels isn't just bad for the environment. You could end up with a huge coal-fired chimney amidst the rolling hills of your backyard.

There are several new approaches to electricity generation currently being explored. One of the most promising is geothermal energy. This involves drilling down into the Earth's crust and using the natural heat found below to power turbines, or to create steam that drives generators.

Another exciting development is the use of ocean waves and tidal currents to generate electrical power. Wave energy has been around for some time, but recent advancements in technology are making it more feasible to capture this energy and turn it into an economical source of electricity.

Finally, research is being done on ways to use solar energy more efficiently, such as through photovoltaic cells and artificial photosynthesis. All of these technologies will help to reduce our dependence on fossil fuels in the future.

Yes, one type of solar energy technology that is showing the most promise for the future is photovoltaic (PV) cells. PV cells are made from semiconductor materials and they use the energy from sunlight to generate electricity. The advantages of this technology is that it is clean and renewable. It does not require any fuel or produce any emissions, making it an attractive option for reducing our dependence on fossil fuels. Additionally, PV cells are becoming increasingly efficient at capturing the sun's energy, meaning that more electricity can be generated in a given space than ever before. Furthermore, advancements in PV cell construction have allowed for decreasing costs, thus making them more accessible to the public. Finally, PV cells have been integrated into many other technologies, such as vehicle charging and building-integrated photovoltaics (BIPV), further expanding the possible applications of this technology and increasing its potential impact on the future of energy.

PV cells are attractive for reducing dependence on fossil fuels because they do not require any fuel or produce any emissions. This means that by using PV cells we can generate electricity without any of the environmental damage associated with burning fossil fuels, such as air and water pollution. Furthermore, PV cells can provide a more secure source of energy since they are renewable, meaning that they can continue to produce energy indefinitely if properly maintained.

Another article,, 7/31/23
https://www.benzinga.com/pressreleases/23/07/ab33460123/
schneider-electric-and-pacific-gas-and-electric-company-announce
-deployment-of-distributed-energy

*https://www.theguardian.com/australia-news/2023/jul/31/
australian-electric-vehicles-ev-sales-rise-increase*

Compressed Air

Now, here's a theory that probably sprung up out of a pizza parlor in DC where conspiracy theories thrive!

Compressed Air Energy Storage **(CAES)** is a system that uses excess electricity to compress air and then store it, usually in an underground cavern. To produce electricity, the compressed air is released and used to

drive a turbine. In a typical CAES design, the compressed air is used to run the compressor of a gas turbine, which saves about 2/3 of the energy needed to operate the turbine.

This leads to a reduction in natural gas consumption and can cut carbon dioxide emissions by 40 to 60 percent depending on the design. **CAES systems** have a large power rating, high storage capacity, and long lifetime. However, because CAES plants require an underground reservoir, there are limited suitable locations for them.

Only two commercial CAES plants exist in the world today, located in Germany and Alabama.

Compressed air energy storage (CAES) is a form of energy storage that uses compressed air to store and retrieve energy. By compressing the air in tanks or cavities and storing it until it is needed, CAES can provide ancillary services such as frequency regulation and black start capabilities, making it an attractive option for utilities.

In addition, CAES can power roller coasters and other amusement park rides while still being sustainable. This is because when the stored air is released and converted back into electrical energy, the same air is recycled and reused, reducing the amount of energy wasted. Furthermore, the process of compressing the air is often powered by renewable sources such as wind or solar, making the CAES process overall more sustainable than other forms of energy storage. Finally, its portability makes it an even more efficient and reliable source of energy. Overall, compressed air energy storage is an efficient and sustainable form of energy storage that can provide numerous benefits, including powering roller coasters and other amusement park rides.

Have you ever read that **wind energy is the cheapest** energy source available? Now, please let me give you the rest of the story,,,

These public claims are based on cost estimates that assume the lifespan of wind turbines to be 30 years. However, according to the U.S. Energy Information Administration, the lifespan of wind turbines is about 20 to 25 years.

Actual data on installations is demonstrating far different data. MidAmerican Energy, which now operates more than 3400 in 37 different projects in Iowa, show wind turbines are reaching the end of their lives even faster. At one point they were forced to re-power turbines merely 14 years after they were installed. Current histories of existing coal, nuclear, natural gas and hydroelectric plants are demonstrating that they can generate electricity for more than 50 years.

MidAmerican Energy has invested $14 billion in wind projects.

By not factoring in this additional spending, these reports not only underestimate the true cost of wind energy, but overestimate the cost of power plants capable of generating electricity for more than 30 years.

In other words, these cost-estimates tell us that wind energy would be the cheapest source of energy if all power sources produced electricity for a similar period of time ... but they don't!!!

Additionally, because wind turbines can only produce energy when the wind is blowing, they generate electricity less frequently than other generation sources.

In Minnesota, wind farms produced electricity only 34.67% of the time in recent studies.

This is far lower than coal at 56.9%, hydropower at 64% and nuclear at 84.6%. Natural gas plants are utilized at a lower rate of 17.6%, but this is because natural gas are used as a "backup" source of electricity when the wind isn't blowing, (an additional cost that is not included in the cost of wind turbines.)

About MidAmerican Energy:

MidAmerican Energy, headquartered in Des Moines, Iowa, serves 813,000 electric customers in Iowa, Illinois and South Dakota, Information about MidAmerican Energy is available at midamericanenergy.com and company social media channels.

https://www.crowrivermedia.com/limited-lifespans-of-wind-turbines-result-in-higher-energy-costs/article_ce7a2c93-0e13-586c-aa8e-cbd8a826c76a.html

Wind farm photos that missed their day of glory on social media channels,,, As stated, the replacement rate is faster than expected and given the current very high recycling costs, there is a real danger that all components will go directly into landfill.

Wind energy is experiencing a boom, but in a pattern eerily reminiscent of the Pennsylvania oil boom in the early 1900's, wind farms are building ever larger turbines to farm wind energy further and further from shore. This trend carries risks, especially as turbines come with largely hidden costs. Increasing evidence suggests that although larger turbines can capture more energy, at a certain point the costs of maintaining and decommissioning large turbines located far offshore will outweigh the benefits of that energy capture. If wind farm operators are to avoid creating an environmental and economic disaster in the longer term, they need to begin factoring realistic maintenance and decommissioning costs into their projections.

Photos licensed under the Creative Commons Attribution-Share Alike 2.0 Generic license.

Here comes the storm!

Many people think lighting is harmful, but, Benny Franklin disproved this theory. Here is an example of a power company in Tennessee recharging the Grid...

Electron Harvesting From Storms:

Here is your last option. In another new development, we will soon be seeing electron harvesting from snow.

Skiing enthusiasts have discovered that dust particles are electrically charged thereby attracting moisture. When this is measured in the winter, the cold air turns the moisture into snow and these flakes actually carry an electrical charge. Electricity forming in storm clouds comes about when the air is moving vertically up and down causing the moisture condensation to release its charge.

Water molecules align differently under electric fields and attach to dust particles at different temperatures. This is why a humid day does not produce any electrical charge, but storms with violent air movements do.

This is a seasonal phenomenon. In the summer, we have lighting, in the winter we have snow. One logical storage solution is easy in the winter—piles of snow

piled up along side of the road or parking lot., or on the picnic table out on the veranda. In the summer, we have to be much more creative, like flying a kite.

It must be noted that noted that there is a fine balance between increased air pollution, dust and temperature that must be met storms to produce snow.

And, once again, this source is not reliable.

Here is a recap of some of the data shared on previous pages… *Just what you really wanted,,,*

5 different types of energy storages…
with efficiency from 25% to 90%

The energy sector has always been associated with engineering advancements. This will continue.

Lithium-ion batteries. "Fresh", agile, expensive Energy storage in lithium-ion batteries is considered one of the most efficient. But only until the battery begins to degrade.
Pros: fast construction, almost instantaneous output of the stored energy (tenths of a second)
Cons: price, degradation, absence of disposal, acquisition of rare earth elements
Li-ion 88 – 90%

One of the main problems with lithium-ion systems is that they degrade over time. The degree of degradation depends on the intensity of the battery use – how often it worked during peak loads, how often it was discharged to zero. But on average, after ten years, the battery capacity decreases to "economically disadvantageous" levels (it applies to industrial facilities) – 50 of the declared for new batteries. Less capacity is less efficiency. Degradation results another problem with lithium-ion batteries – their disposal.

Power-to-gas. Gas storages: innovative, "green", less efficient
 Pros: Possible to supply methane to container transportation
Cons: low efficiency, expensive electrolysis process
Power-to-gas 50%

PSP Pumped storage power plants are not as fresh as lithium-ion batteries or hydrogen storages.
PSP(new) 75%
Pros: inexpensive storage, high capacities
Cons: expensive and lengthy construction, the need for a suitable landscape

Gravity energy storage systems. Unusual, cheaper. The principle of work is based on gravity and friction, basically, it is similar to the PSP.
Gravity storages 80 - 85%
Pros: fast construction and comparative low cost
Cons: new technology with uncertain efficiency

Thermal energy storage. Or a heated "philosophical" stone by Siemens
Pros: inexpensive
Cons: Not very efficient, only suitable for seasonal energy storage
Thermal storages 25%

https://kosatka.media/en/category/blog/news/5-sposobov-hraneniya-energii-i-naskolko-oni-effektivny#:~:text=Energy%20storage%20in%20lithium%2Dion,one%20of%20the%20most%20efficient.

Tesla - Toasters - Teslas

You have had this opportunity to see a few of the ways Eddie Current has enriched our lives and, also, a few of the media hypes on supposed abuses, most of which are deep dives into misinformation and hidden agendas. I am confident that you will see this publication as a contribution to point out complexity, involvement, and dependency on our need for power.

Hopefully, our ingenuity will eventually recognize the many options and opportunities we have to combat Climate Change and governments will not have to tax us into an acceptance of misinformed beliefs.

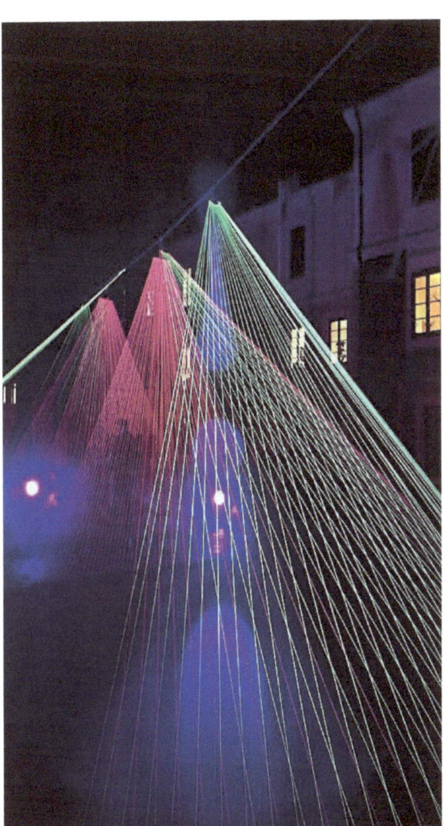

LASERS:

In the past and still today, entertainers use lasers for their colorful visual effects, and doctors use them to perform delicate operation, often to the human eyeball. Also, as an electronic appliance, Lasers can be so powerful that they can vaporize a bulldozer 2000 yards away.

Instruction manuals needed to be rewritten to include instructions to change the power setting from "Vaporize Bulldozer" to "Delicate-Eye Surgery" depending on the task at hand.

The END

Additional References: **Disclaimer —- AI was NOT used for any part of this story!**

The IEA's Digital Demand-Driven Electricity Networks (3DEN) Initiative and IEA - 4E TCP Electronic Devices and Networks Annex are co-organizing this webinar illustrate how blockchains work, how cryptocurrencies and other applications employ the blockchain, and why the resulting energy consumption is so high. A potential solution to significantly reduce the energy consumption used for validation will also be presented.

The webinar will draw from the key findings of the report *Blockchain Energy Consumption - An Exploratory Study.* The report was commissioned by the Swiss Federal Office of Energy and authored by Mr Vlad Coroama, Independent researcher, Roegen Centre for Sustainability (formerly at ETH Zürich).. The report contains a detailed analysis of blockchain, its use in cryptocurrencies and beyond, and the resulting energy consumption.

Based on the Swiss report, the IEA-4E Electronic Devices and Networks Annex has published a policy brief on this topic, aimed at policy makers.

The IEA gratefully acknowledges the Italian Ministry for Ecological Transition for their support for this webinar as part of their contributions to IEA's Digital Demand Driven Electricity Networks (3DEN) Initiative on power system modernization and effective utilization of demand side resources through digitalization and to the Clean Energy Transitions Programme. The IEA equally acknowledges the IEA-4E Electronic Devices and Networks Annex sponsorship of this webinar.

Select Works Of Dr. Willie Soon:

"Was the 20th Century Climate Unusual? Exploring the Lessons and Limits of Climate History," Dr. Willie Soon, 5/16/2003

"Lessons & Limits of Climate History: Was the 20th Century Climate Unusual?," Dr. Willie Soon and Dr. Sallie Baliunas, April 17, 2003
"Reconstructing Climatic and Environmental Changes of the Past 1000 Years: A Reappraisal," Dr. Willie Soon, Dr. Sallie Baliunas, Sherwood B. Idso, Craig Idso and Dr. David R. Legates, April 11, 2003
"Extreme Weather Events: Examining Causes and Responses," Dr. Sallie Baliunas and Dr. Willie Soon, 3/25/2003
"Climate History and the Sun," Dr. Sallie Baliunas and Dr. Willie Soon, June 5, 2001
"First Eurocongress on the Solar Cycle and Terrestrial Climate," Dr. Willie Soon, Dr. Sallie Baliunas, Kirill Ya. Kondratyev, Sherwood B. Idso and Eric S. Posmentier, September 30, 2000
"Calculating the Climatic Impacts of Increased CO2: The Issue of Model Validation," Dr. Willie Soon, Dr. Sallie Baliunas, Kirill Ya. Kondratyev, Sherwood B. Idso and Eric S. Posmentier, September 30, 2000
"Comments on New Danish Solar Study," Dr. Willie Soon, April 2, 2000
"The Sun Also Warms," Dr. Sallie Baliunas and Dr. Willie Soon, March 24, 2000
"Solar Variability and Climate Change," Dr. Willie Soon, January 10, 2000
"Increasing Carbon Dioxide and Global Climate Change," Dr. Sallie Baliunas and Dr. Willie Soon, January 1, 2000
"A Scientific Discussion of Climate Change - Comments on," Dr. Sallie Baliunas and Dr. Willie Soon, 11/1/1997

Additional References:

High electricity prices have Europe facing deindustrialization; don't let it happen here | The Hill OPINION-ENERGY AND ENVIRONMENT—Jan 2024

The Shock Doctrine
The Shock Doctrine: The Rise of Disaster Capitalism is a 2007 book by the Canadian author <u>Naomi Klein</u>.

Detailed reference data:

https://m.youtube.com/watch?v=pr0LkPMZ-qc

https://youtu.be/QsiGiziM7Vw?si=7ubshkj3YmQHXpRo

https://www.smithsonianmag.com/smart-news/honeybee-swarms-can-produce-as-much-electric-charge-as-a-thunderstorm-180981005/

https://youtu.be/QsiGiziM7Vw?si=7ubshkj3YmQHXpRo

https://www.pcmag.com/how-to/what-is-mpge-the-new-fuel-efficiency-rating-for-hybrids-and-evs-explained

https://kosatka.media/en/category/blog/news/5-sposobov-hraneniya-energii-i-naskolko-oni-effektivny#:~:text=Energy%20storage%20in%20lithium%2Dion,one%20of%20the%20most%20efficient.